Information Systems Engineering and Management

Volume 1

Beatriz Graça
DEGEIT
University of Aveiro
Aveiro, Portugal

Manuel Au-Yong-Oliveira
INESC TEC, GOVCOPP, DEGEIT
University of Aveiro
Aveiro, Portugal

ISSN 3004-958X ISSN 3004-9598 (electronic)
Information Systems Engineering and Management
ISBN 978-3-031-59913-2 ISBN 978-3-031-59914-9 (eBook)
https://doi.org/10.1007/978-3-031-59914-9

This Springer imprint is published by the registered company Springer Nature Switzerland AG
The registered company address is: Gewerbestrasse 11, 6330 Cham, Switzerland

Foreword by Emídio Gomes

The experience presented by Beatriz Graça and Manuel Au-Yong Oliveira leads to put in question the reasons behind our conformity in everyday life. It challenges the status quo of many organizations, which often focus on the negative aspects of interpersonal relationships in their form and functioning, with a modus operandi that feeds on internal information circuits, a common methodology used by a lot of leaders during their daily power management. This book even sometimes reminds us of the moments of reading Scott Adams' forced caricatures and his postulation of the Dilbert principle.

The most efficient and developed countries clearly have fewer negative organizations, especially in their public institutions, where the established culture is more conducive to the growth of these factors. However, it is essentially in the world of private and open economy that the difference is clearly accentuated. The present book, through a proposal for an experimental approach methodology in case analysis, represents a positive contribution to our understanding of the evolution of societies and their interaction with the real economy.

Emídio Gomes
Full Professor
Porto University
Porto, Portugal

Rector of UTAD
Universidade de Trás-os-Montes e Alto
Douro
Vila Real, Portugal

Foreword by Evelina Coutinho

In today's business context, subject to constant change and new challenges, the concept of competitiveness takes on vital significance. To remain in the market, companies must have the ability to reflect on how organizational culture influences performance, redefining, if necessary, policies and strategies. It is in this context that the book *Competitiveness Strategies for Negative Organizations: Challenging the Status Quo* emerges as a lighthouse for leaders and entrepreneurs.

In a world where positivity is synonymous of happiness, addressing strategies for organizations to navigate negativity and transform into positive organizations is a courageous and necessary endeavor. The book immerses the reader in the complexity of negative organizations, providing a different analysis than the conventional literature. The authors explore the theme from different perspectives, in a fluid and simple way, based on interviews and a focus group interaction. They properly frame each aspect, presenting a range of strategies that lead to the establishment of positive organizations to improve competitiveness, guiding readers in understanding the key factors in the transformation of organizations.

As you embark on this journey, you will find a collection of ideas that challenge preconceived notions and invite you to reconsider the very essence of competitiveness. The book *Competitiveness Strategies for Negative Organizations: Challenging the Status Quo* is a manifesto for change, a scream for those who pursue innovation, growth, and sustainability. This book serves as a guide for leaders, managers, and students, who seek to navigate the uncharted waters of negative organizations, turning challenges into opportunities and forging a path to a future where resilience and competitiveness go hand-in-hand. May this book inspire, challenge, and unleash a new era of strategic thinking that empowers organizations to redefine and rewrite their future and the rules of competitiveness.

Evelina Coutinho
Senior Executive in Technology
and Innovation Management
and Researcher with
GOVCOPP—Research Unit
in Governance, Competitiveness
and Public Policies
University of Aveiro
Aveiro, Portugal

Foreword by José Magano

I am pleased to introduce *Competitive Strategies for Negative Organizations: Challenging the Status Quo*, a new tangible result of the authors' extensive research on negative organizations. The book not only looks at what determines the competitiveness of negative organizations, but also identifies strategies for dealing with it. In addition, the chapters include important results from a mixed-methods research, describing different cases focusing on Portugal and Google, which makes the text interesting in its own right.

Not surprisingly, the research documented in this book acknowledges the crucial role of leaders and organizational culture in shaping the (negative) nature of the organization and its competitiveness and discusses the organizational conditions that allow negative leadership to develop and persist in the workplace. Indeed, what emerges from the reading is the importance of leadership, approached from various perspectives. Every reader will be able to identify situations reported in the text, namely, the encouragement of the use of lies, manipulation, and abuse, typical of a negative organization and a toxic leadership style. A better understanding of the negative aspects of leadership will certainly help organizations to better address the ability of their leaders to lead effectively. The research findings on negative leadership should help organizations to avoid selecting negative leaders, identify the negative leaders they may already have, and create the conditions for negative leaders to become better leaders. This often involves a change in management towards a different mindset—one driven by leaders who not only understand the challenges of a rapidly changing marketplace and the determinants of competitiveness, anticipate, adapt, and drive profitable change, but who are also able to motivate and empower employees to become (and feel) valuable organizational assets who play a pivotal role as innovators.

This book can be read in one breath. So do not delay: broaden your horizons by reading *Competitive Strategies for Negative Organizations: Challenging the Status Quo*, after which you will indeed switch to this exciting view of organizations.

José Magano
Entrepreneur and Senior Lecturer
and Researcher—Research Centre
in Economics and Business Sciences
University of Aveiro
Aveiro, Portugal

Preface

The topic under analysis—Negative Organizations—is still under studied in the literature. The research question—to identify the problems that influence the competitiveness of negative organizations—aims to understand what a negative organization consists of, but also to establish strategies for its transition to a positive organization. This research used a PRISMA methodology (The Preferred Reporting Items for Systematic reviews and Meta-Analyses) to analyze the literature and thus deliver the results in a clear and transparent manner. Subsequently, six interviews and a Focus Group were conducted in order to understand the perspective of the selected purposive samples. The results revealed that the leaders and managers play a central role in an organization and therefore have the power to transform it into a negative or positive organization, depending on their approach in various dimensions, such as made evident in the proposed FLOSE model (which incorporates five main factors including financial resources and stability, leadership style, organizational culture, the strength of the *status quo*, and the type of external environment). In sum, is creativity and innovation stimulated? How does the firm view change? How are promotions decided within the organization? This research is intended to serve as a tool for managers and leaders to use in their day-to-day work in order to identify negative organizations or negative dimensions of the organization and to develop a strategic plan for the transition to a positive organization.

Aveiro, Portugal

Beatriz Graça
Manuel Au-Yong-Oliveira

Acknowledgements

My deepest thanks go to all those involved in this study, without whom this book would not be complete.

I would also like to thank my coauthor, Prof. Manuel Au-Yong Oliveira, for his advice and inspiration on the subject under study. Thank you, Professor, for all your help!

I would also like to thank Prof. Frank S. Fialho for the inspiration he has given me over the last few months. Thank you for the inspiration and devotion you have given me in your Strategic Marketing classes.

A very special thank you to my family, especially my parents and siblings, for their unconditional support.

Thank you all!

—Beatriz Graça

I would like to thank all of my current and previous colleagues for all we have shared while on the job and in the field. It has been a fantastic learning experience. My thanks also to my students, for whom I work daily, and to my family and friends for their constant support and for being my safe haven.

São, João, Benedita, Joana and Leonor, *muito obrigado.*

—Manuel Au-Yong-Oliveira

Contents

List of Figures

List of Tables

Chapter 1
Introduction

Abstract This chapter introduces the study and provides definitions of some central concepts. What constitutes a positive culture and organization? What constitutes a negative culture and organisation? Related concepts, such as competitiveness, leadership, the role of senior management, and the civil service versus private enterprise—as well as examples of the concepts—are also discussed. The background for the book is set herein.

Workplace culture involves the values and belief systems, the attitudes and set of assumptions, shared by all employees in an organization, and often not readily visible or easily comprehensible by outsiders. Indeed, it is a mode of functioning unique to the organisation which "comes naturally" and dictates how insiders should behave in certain situations—without having to seek counsel from above. In fact, in practice, organisational culture encompasses various dimensions of the employer-employee relationship regarding workplace norms and operating procedures [1]. This culture plays a central role in influencing and encompassing the other dimensions of an organisation, such as the degree of innovation and its competitiveness, among others.

It would be expected, given the central role of organisational culture, that, as with other aspects of an organisation, it could lean towards a more positive or more negative culture, as follows:

1. A positive culture and organization—a culture that accepts and promotes innovation/innovative people and the best sustainable organisational strategy possible, irrespective of the *status quo* or existing hierarchy; and
2. A negative culture and organization—a culture that promotes relationships and their continuation (however bad they may be for the organisation) while expelling or inhibiting innovative human resources due to those existing relationships (of which the good human resources are not a part) and the *status quo* (existing hierarchy) of the organisation. The best and most sustainable organisational strategy is not followed.

B. Graça and M. Au-Yong-Oliveira, *Competitiveness Strategies for Negative Organizations*, Information Systems Engineering and Management 1,
https://doi.org/10.1007/978-3-031-59914-9_1

Obviously, the organisational culture of an organisation cannot be the only criterion for measuring the positivity or negativity of an organisation, an aspect that will be explored in greater detail throughout this book.

Competitiveness is another criterion for measuring an organisation's positivity, as it reveals, for reasons explored later, an organisation's reluctance or openness to adapting to the market in which it operates, adding monetary value to its stakeholders. After all, all organisations work and exist to be competitive and, consequently, to generate a financial return. However, financial return cannot be an organisation's main focus, since this aspect, when combined with a commitment to innovation and competitiveness within an organisation, is a natural consequence.

The leadership of an organisation is another dimension for measuring the positivity or negativity of the organisation, since it is the leaders (and managers) who foster and encourage change within the organisation. However, there is a danger that the organisation's leadership will steer its growth in an uninnovative and therefore uncompetitive direction, running the risk of bankrupting the organisation. For this reason, leaders must take account of market changes and have the courage and receptivity to make the necessary changes to the organisation's structure in order for it to succeed, over time.

In addition, the government of each country also plays a key role in influencing the existence of more positive or negative organisations. They should therefore play an active role in encouraging organisations to be more positive and competitive in the global market.

The main aim of this book is therefore to define the concept of negative organisations and to identify strategies that help organisations and their stakeholders, such as top and middle managers, employees and investors, to make the working environment more positive and the organisation more competitive. To this end, a qualitative methodology was used, through six interviews and a focus group, the results of which will be presented later.

It is important to note that, while much has been discussed about how organisations can act to become more positive, it is equally important to understand what characterises a negative organisation and how it can be transformed into a positive one, even though this is a relatively recent concept and one that is little covered in the literature.

Therefore, this research seeks to comprehensively explore and analyse the impact of organisations and their organisational culture on competitiveness, with the aim of providing insights and strategies that can help companies adapt to contemporary challenges and stand out in the market.

If a senior manager seeks, for example, due to a professional and informal alliance with a colleague, and/or due to a friendship tie with them, to protect and promote this colleague who is aggressive, untruthful and a bully—as they act in their own interest, to promote themselves in detriment of honest and more innovative/valuable employees, with whom they are competing for a promotion, then we are witnessing the case of a negative organisation. If this negative informal alliance prevails then the firm may lose valuable human resources (either expelled from the organisation or choosing to leave due to the persecution experienced while on the job) and in the medium- to long run they may fail.

Note that in the civil service failing outright is more difficult and so the case may be distinct (e.g., more serious) than in privately owned firms. In privately owned firms expelling certain human resources is easier than in the civil service (where this may be close to impossible). Hence, in the civil service, we may see more negative organisations, and this is certainly in line with their performance… as many or most publicly-owned firms are not profitable in the sense that they do not generate superior value versus their cost of operation.

Reference

1. Hsieh, N.H., Klopfenstein, A., Mehta, S.:. Culture at Google. Harvard Business Review (2020)

Chapter 2
Literature Review

Abstract This chapter is a literature review, namely performed on the Scopus database, which is a major database for the social sciences and hence related to this work. A main theme studied is organizational culture—which is linked to negative organisations, bullying and corporate psychopaths; as well as to positive organisations. The future of organisations is also discussed, as is the importance of innovation, the role of leadership and the role of government. The importance of a literature review is twofold. On the one hand it aims to show what themes are related to the study. On the other hand, it aims to show where certain gaps in the literature exist and which the study may seek to fill. The theme negative organisations has been understudied to date.

The main aim of this chapter is to address issues pertinent to this study—*Competitiveness Strategies for Negative Organisations: Challenging the Status Quo*. In this sense, it will address topics such as:

i. Organisational Culture;
ii. Negative Organisations, *Bullying* and Corporate Psychopaths;
iii. Positive Organisations;
iv. The future of organisations, the importance of innovation, the role of leadership and the role of government.

When it comes to analysing the positivity or negativity of an organisation, it is necessary to have two key concepts as a basis, which serve as a foundation and which, when combined with the other points that will be covered throughout this research project, make it possible to carry out a preliminary analysis of the negativity and positivity of a company. These concepts are (i) the organisational structure and (ii) the meritocratic system.

It is therefore important to begin this discussion by addressing the issue of how organisations are structured and how decisions are made on a day-to-day basis. This aspect is fundamental to identifying how organisational structure contributes to organisations achieving their purpose, i.e., the objectives previously outlined [1].

B. Graça and M. Au-Yong-Oliveira, *Competitiveness Strategies for Negative Organizations*, Information Systems Engineering and Management 1,
https://doi.org/10.1007/978-3-031-59914-9_2

Defining the structure of the organisation thus makes it possible to clarify the necessary functions, relationships and authority [2] and, consequently, influences the way managers work, as well as the decision-making process [3]. It is therefore essential that the structure is aligned with the organisation's strategy in order to achieve the desired results [3] and thus ensure that the organisation is able to keep up with any paradigm shifts that may occur in the various industries [2].

Anticipating these paradigm shifts requires the organisation to have, at all levels of the company and for each function performed within it, the right person with the necessary skills for the job assigned to them. These types of organisations are called "meritocratic" and therefore have the best people possible performing the right roles in the right job for them [2] (versus having the wrong people perhaps even in key positions in an enterprise in order to pay for favours or to gain access to favours). Meritocratic systems allow (i) the quality of the goods (i.e., products or services) produced to be higher, (ii) decisions to be made more quickly and (iii) the resources used to be allocated more efficiently and effectively [2].

On the other hand, in meritocratic organisations, employees are recognised and rewarded solely on the basis of their merit [2]. Purely meritocratic organisations are rare, however, as organisations usually adopt hybrid systems, i.e., they reward their employees not only on the basis of merit, but also on the basis of aspects such as seniority, nepotism, democracy and favouritism [2]. Whether in a business or government context, employees should perform duties that fit their skill level and competencies and are realistically challenging as regards providing them with an opportunity to add value, thereby contributing to the organisation's purpose and growth. The organisation, in turn, should be willing to support them and provide them with opportunities to develop their potential, so that they can make an even greater contribution in the future [2].

In short, a meritocratic system is a very efficient and effective way for an organisation to produce its products and/or services, and the leadership ranks must have the courage and impartiality to implement meritocracies. Consequently, it is important to analyse how the organisation's own culture can influence, positively or negatively, the implementation of a meritocratic system, in which employees are rewarded on the basis of their work and effort.

In a country like Portugal, where professional rewards such as promotions, salary increases or incentives may essentially be based on relationships and contacts [4], choosing the best person to do the job is also naturally influenced by this organisational culture [5, 6]. In particularly corrupt (or negative) organisations, the choice of the person for the job in question is made for the sake of future or immediate benefits for the contractor and the hired or just because of the family or friendship relationship between the parties. As you might expect, choices made on the basis of interpersonal relationships and not on the basis of the person's suitability for the position can obviously harm the sustainability and competitiveness of the organisation, since the most suitable person for the position is not necessarily the one chosen. In this sense, corruption, i.e., the practice of actions that benefit the individual doing them or their family and friends [7], harms the sustainability and competitiveness of the organisation. Choosing the right person translates into better qualifications for making decisions relevant to the job in question. An inappropriate choice can have unpredictable consequences for the organisation which, in the medium to long term, can be detrimental to competitiveness, especially in highly competitive environments [2].

2.1 Organizational Culture

Organisational culture is a fundamental topic in the study of organisations, as it significantly influences the behaviour of individuals, internal dynamics and company performance. When analysing organisational culture, it is important to consider that organisations are made up of individuals from different countries, each with their own culture. A country's culture encompasses beliefs, values, traditions, customs and behaviours shared by its inhabitants, and these characteristics can vary significantly from one nation to another (for example, being more relationship- or more task-oriented [4]). Therefore, understanding the cultural differences between countries is essential to understanding the dynamics of organisational culture. By exploring cultural nuances, it is possible to identify how these differences can influence the practices and values adopted in organisations, as well as the expectations and behaviours of the individuals who are part of them [8].

Organisational culture is a very important aspect for understanding how individuals behave in the organisational environment, thus affecting the strategic decisions of its leadership and the way they work internally. Organisational culture is understood to be the complete set of ideologies, symbols and fundamental values that are shared by the entire organisation, including employees and other *stakeholders*, influencing the way in which the organisation carries out its business [3].

With this in mind, organisational culture has the power to influence and motivate the future of the organisation. As you might expect, organisational culture is closely linked to its long-term competitiveness, as it determines whether the organisation is strategically aligned to meet its objectives and thus be successful in the market it is in, and therefore be competitive.

This competitiveness is also intrinsically linked to innovation and the ability of organisations to position themselves "ahead of the curve", since through innovation organisations achieve improvements in terms of products, manufacturing processes and yields, which translates into a reduction in costs and an increase in the satisfaction of customers and other stakeholders involved. There are countless examples of organisations that have stagnated in terms of innovation, which has led to them losing their position in their market sector or, in the worst case, going bankrupt and disappearing [9]. A truly innovative company will be able to anticipate paradigm shifts and thus remain competitive in its market. There are, however, organisational cultures that prevent the anticipation of paradigm shifts in the market by discrediting innovative individuals who have the necessary skills to change (or maintain) the organisation's path for the better, making it more competitive. In fact, even if there is a high degree of innovation within the lower levels of the company, it is ultimately the decision of the higher-ups that drives it forward, which results in the establishment of organisational patterns that could be extremely damaging to the company in the long run [10].

It is important to note that a paradigm shift does not necessarily translate into gains for the organisation, hence the resistance of many organisations' hierarchies to taking steps to adapt to these changes. However, a truly competitive organisation will be able to discern the changes that will have a positive impact on its competitiveness and that accept rather than discredit these innovative individuals.

The stagnation of organisations in the context of innovation is a complex challenge that can be aggravated by a lack of incentive for creativity and the exploration of new ideas. In addition to organisational cultures that discredit innovative individuals, there are also those that tend to value maintaining the *status quo* and resist any form of disruptive change.

2.1.1 Negative Organizations

As established in the previous chapter, the organisational culture, disseminated by top management and shared by all employees and other *stakeholders*, has the power to determine the organisation's competitiveness. It is crucial that the organisation's leadership adapts to and encourages innovation and the changes that come with it, by developing new ideas and tools that facilitate the implementation of these innovations [11]. This process of developing ideas and creating the tools needed to implement innovation often involves a "transformation process" for the company, which will have to adapt to new processes, new ways of working and a new organisational culture. As you might expect, this is a lengthy and costly process, which is why many Portuguese organisations are unwilling to start it/put it in motion.

Although many of the ideas and tools that initiate this transformation process may come from employees at the bottom of the organisational structure, it is mostly managers who have the necessary power when it comes to implementing them and are therefore indispensable in this process. It is these managers who must realise that implementing innovation can lead to high financial returns in the long term, and whose implementation benefits their organisation in the long term [5, 6]. The leadership of organisations, i.e., executives (senior leaders), entrepreneurs (business leaders) and change agents/innovation champions [12], therefore play a vital role in the transformation process, as they are responsible for outlining well-defined strategies that remove the negativity from this process [13].

In addition, organisations do not live in isolation and are therefore part of a dynamic ecosystem that influences the various organisations within it, i.e., all innovations will directly or indirectly affect the way the organisation operates. It is therefore essential that the organisation evolves while the ecosystem evolves too [11]. Even though organisations know that they can gain many benefits from implementing innovations, it is still difficult to transform themselves. It is therefore important to question the following [13]:

(i) Faced with radical technological advances and market disruptions for various products, why are organisations still so cautious and conservative when it comes to implementing innovations?
(ii) Why aren't organisational executives willing to implement innovations, even though most of them admit that successful innovation will benefit the organisation in the long term?
(iii) How is it that just one dissenting colleague, a single voice of negativity in a business unit or project team that is optimistic and enthusiastic, can cast such a long shadow over the morale of the group? How does one "bad apple" always seem to spoil the whole bunch?

Is the answer to the above three questions one of the steps needed to make companies less negative? But what makes an organisation negative? This concept covers several dimensions, so there is no single definition. It is therefore an "umbrella concept", which encompasses dimensions such as (i) Creativity, (ii) Corruption, (iii) Innovation, (iv) Aversion to change, (v) Poor internal environment, (vi) Privilege for relationships and (vii) Negative management of the organisation's strategy. The following table (Table 2.1) shows in full how these elements can lead an organisation to be negative.

All the characteristics listed in Table 2.1 have a common denominator, i.e., they all affect the organisation's continued competitiveness. At the opposite end of the spectrum, positive organisations (a topic developed in greater depth in Sect. 2.1.2.) are organisations that are sustainable and competitive in the long term, whereby there is harmony between all the organisation's employees, including those in different organisational hierarchical positions [5].

From the dimensions listed in Table 2.1, it can also be concluded that negative organisations have the potential to destroy their own competitiveness. Companies like Nokia and Kodak assumed that they did not need to be innovative because their products were insurmountable and would always be successful. As expected, this was not the reality, because in a world of transient advantages, organisations have to be quick to notice possible changes that could affect their whole business [20]. When they are unable to accept and keep up with radical paradigm shifts, they run the risk of failing and sometimes even disappearing, as was the case with these two companies, regarding their main business units. Given the resistance to change inherent in each individual, managers sometimes deny that such changes mean anything positive for the organisation. In order to remain competitive, it is critical that those in management and leadership positions within the organisation are able to (i) interpret patterns in the evolution of the state of things and (ii) respect the truth even if it has inconvenient or controversial implications [5].

On the other hand, the role of employees cannot be overlooked either. Although the negativity of an organisation is largely linked to top management, it is the employees who spread the organisational culture instilled in them by their bosses. This role is also a dimension that affects the competitiveness of each organisation.

In short, employees and managers, i.e., the organisation's human resources, are one of the keys to an organisation's success, but their attitudes can contribute to a

Table 2.1 Definition of the concept of negative organisations (own elaboration)

Defining the concept of negative organisations	Authors
All employees in an organisation, including top management, are in organisations to serve a very specific purpose, in exchange for a reward (i.e., salary, skills creation and experience). However, **negative organisations seek to perpetuate relationships and personal interests, serving only a few people in power**, rather than seeking the continued competitiveness of the organisation and adding value in all the tasks that each person in the organisation performs	Au-Yong-Oliveira [5, 6]
Employees continually use unethical, anti-competitive, deviant and illegal practices in order to achieve their goals, thus becoming corrupt. Corruption is traditionally defined as the adoption of illegal activities that result in obvious gains in power and influence, directly or indirectly, which seems to be the goal of the employees who practise it. It can also be seen that corrupt employees encourage others to participate in and ignore their malpractice. Corruption is **thus a major threat to the sustainable future of organisations** and is therefore one of the dimensions of a negative organisation	Vicker [14], Au-Yong-Oliveira [5, 6]
In **negative organisations there is no room for creativity**. In these organisations, creativity is usually limited to the marketing and R&D department (if this is a department that is already in place), and there is no room for its implementation in other sectors of the organisation, even if they can bring benefits to the whole organisation	Amabile [15]
As already mentioned, resistance to change is an instinct intrinsic to most individuals, but this instinct needs to be counteracted. As well as **being resistant to internal changes, negative organisations are also not aware of external changes** (i.e., changes in customer consumption habits, regulatory changes, changes in the industry, among others), which means that the necessary measures are not taken to take these changes into account, thus affecting their competitiveness. The top management of a negative organisation is unable to communicate change effectively in order to align all of the organisation's stakeholders	Au-Yong-Oliveira et al. [9] Macdonald et al. [2] Schein [16]
There is no strategy without innovation and an organisation's competitiveness depends essentially on its ability to change its procedures and products/services in line with the changes taking place in its ecosystem. Innovation is therefore essential, yet in **negative organisations innovative individuals are not recognised and may even be discredited**	Au-Yong-Oliveira [5, 6]
Negative organisations choose to be short-sighted, in other words, they choose to ignore changes that will be inevitable in the future. On several occasions, the stubbornness of managers has led to the loss of market leadership by insisting on products that, in the near future, would no longer be competitive	Au-Yong-Oliveira et al. [9]

(continued)

Table 2.1 (continued)

Defining the concept of negative organisations	Authors
Employee rewards go beyond monetary rewards, although these are also important, and the organisation needs to instill in its employees a sense of pride in their day-to-day work while achieving their own goals. Given the lack of rewards and employee pride in **negative organisations, they are characterised by problems related to absenteeism, employee turnover, the quality of their final product/service and meeting deadlines**	Oliveira [17] Tupper and Ellis [18]
In a negative organisation, individuals who do not deserve it are often rewarded on the basis of the interests of certain individuals. It turns out that, due to their skills or lack of them, these individuals are not the best option for the organisation's competitiveness	Au-Yong-Oliveira [5, 6]
The bad atmosphere and power games experienced within the organisation characterise a negative organisation. ***Bullying*** **and corporate psychopaths (a concept better explained in Sect.** 2.1.1.1) **damage a good organisational environment and, consequently, have a negative impact on the organisation's productivity and competitiveness**	Boddy [19]
Negative organisations are also characterised by **leadership that perpetuates poor strategic management** of the organisation and its core business	Au-Yong-Oliveira [5, 6]

decline in the organisation's competitiveness. It is essential that the management team, as well as all other employees, understand that the organisation's success is temporary if innovation and transformation measures are not imposed. The decisions made must always be aimed at the success and competitiveness of the organisation, and when they do not, organisations become negative [5].

2.1.1.1 Bullying and Corporate Psychopaths

The bad organisational environment, bullying and power games experienced within the organisation are essential elements in defining a negative organisation, which is why an organisation's leadership team plays a key role in identifying and resolving these situations.

Workplace bullying, defined as the repeated unethical and unfavourable treatment of one person by another in the workplace, including behaviour designed to belittle others through humiliation, sarcasm, rudeness, overworking an employee, threats and violence [19], causes great emotional pain for the victim in particular, but also has a significant negative impact on the organisation, namely on productivity, leading to large financial costs. Bullying is thus a tactic used (i) to intimidate others and cause fear of confronting the corporate psychopath involved (ii) to humiliate subordinates and (iii) to confuse and disorientate those who may be a threat to the corporate psychopath's activities [19].

Although most bullies engage in this behaviour with the intention of hurting, denigrating, discouraging and damaging the victim, there is a small group of people who use bullying as a tool for advancement and growth within the organization—these individuals are called corporate psychopaths. It should be stressed that not all bullies are corporate psychopaths, however, corporate psychopaths who use bullying as a tool for growth within the organisation are particularly dangerous [19].

It is therefore important to understand what corporate psychopaths are and how they harm the competitiveness of the organisation. It is estimated that 1% of the adult population is psychopathic, which means that they show no feelings of love or empathy for other individuals, and are therefore characterised as being: selfish (looking out for themselves), cold (rather than empathetic and comprehensive of others), calculating (overly rational), ruthless towards others ("anything goes", irrespective of an absence of ethics) and hence representing a high risk to their employing organisations as well as to their own colleagues and citizens in general (external members of society) [10]. The concept of corporate psychopaths comes from combining the term "psychopath" with the term "corporate", which means a psychopath who works and operates in the organisational field [19]. Corporate psychopaths have been identified through testimonies and are linked to negative organisational outcomes in terms of staff abandonment, bullying, conflict and disorder in the workplace.

Given their calculating, ruthless and cold nature, as mentioned by Boddy [19], decision-making by these members of the organisation tends not to take other employees into consideration, who do not seem to work in their favour. However, these individuals can also be very charismatic, charming and eager manipulators, which seems to make them stand out from their peers as potential managers. Since an organisation's culture is shaped and disseminated by all employees, especially those in positions of leadership and authority within the organisation, corporate psychopaths seem to have an advantage here.

The leadership of corporate psychopaths is characterised by lying (such as claiming qualifications that the corporate psychopath does not actually possess), parasitic behaviour and abusive conduct, such as public bullying and manipulation [10]. It can be seen that the corporate psychopath uses bullying, i.e., actions and behaviour that lead to dysfunctional and negative results for the organisation and its individuals, to confuse and disorientate those who may pose a threat to them [19], irrespective of the positive value they create for the organisation or even because of that.

But what are the consequences of bullies and corporate psychopaths for organisations? These individuals are very detrimental to the competitiveness of organisations, since they worsen the job satisfaction of their colleagues, which is directly proportional to the levels of absenteeism and staff turnover. Absenteeism and staff turnover also result in major financial costs for the organisation, especially when it comes to hiring and training new employees. In addition, corporate psychopaths will undermine the good organisational environment, which translates into losses in productivity and motivation. All of this is, as mentioned above, intrinsically linked to the competitiveness of the organisation, and consequently to a continuous and uninterrupted loss of this competitiveness.

In short, a good organisational environment is essential for the full functioning of an organisation, whether public or private, as employees will feel more motivated and consequently have higher levels of productivity. It is therefore a matter of competitiveness to combat these negative individuals and organisational cultures.

2.1.2 Positive Organizations

The previous chapters defined some of the criteria for classifying a negative organisation. It is therefore important, for the sake of comparison, to define the concept of a positive organisation, which is the antithesis of the first.

It is important for individuals to develop a positive relationship with the organisation they work for, as well as with their colleagues, as these relationships ultimately affect their personal lives too.

Creating an organisation that is positive is an arduous process, and the concept itself is subjective and varies from person to person. This means that the positivity of an organisation is governed by different criteria for each individual. For this reason, it is important for senior (and middle) leadership to understand its employees in order to understand how it can mould them to the positivity criteria defined by the organisation [2].

Just like the definition of negative organisations, the concept of a positive organisation is also complex and involves various aspects, which are listed in Table 2.2.

Although many organisations promote themselves as being positive organisations, it can be concluded that, given the complexity of the subject, most companies still have a long way to go [5].

Nevertheless, organisations that take steps to become more positive should also be commended, as all great changes start with a small step in the right direction.

2.2 The Future of Organizations

Nowadays, the world is constantly changing, whether socially, economically or technologically, so it is necessary to understand what is happening at all these levels from different perspectives, as well as realising how these different dimensions are interconnected. As you might expect, all these changes also have an impact on organisations and the way they work.

One of the great driving forces behind these constant changes is the globalisation of products and services, which has taken place in recent years. In addition, globalisation in all its fullness has allowed organisations to gain other perspectives on ways

Table 2.2 Definition of the concept of positive organisations (own elaboration)

Defining the concept of positive organisations	Authors
A positive organisation recognises the work done by its employees, regardless of their hierarchical position. **A positive organisation thus recognises the contribution made by each member of the team,** and by showing genuine support for the work done, employees recognise that the organisation encourages them to develop their professional and personal skills. Furthermore, in a positive organisation, each employee's individual skills match and adapt to the demands of their role	Macdonald et al. [2]
Positive organisations **recognise the importance of "Innovation Champions"**, i.e., individual employees who have a major influence on the dissemination and acceptance of the organisation's innovations. These employees are essential for the whole organisation to be committed to the changes that come with implementing innovations. Thus, in a positive organisation, leaders do not feel threatened by having brilliant employees who have brilliant ideas	Howell [12]
Conflict between individuals is inevitable, yet **positive organisations consider conflict to be positive, constructive and essential for teamwork and organisational effectiveness and, consequently, for ensuring the organisation's competitiveness**. Diversity (of skills, race, sex and gender) between individuals, while adding various challenges, is important for different perspectives to emerge, as well as for developing new procedures The organisation's role is therefore central to ensuring that the challenges resulting from diversity are well managed and recognised, making it an inclusive place for different individuals	Tjosvold [21] Tjosvold and Su [22]
The main goal of all organisations is to be competitive and, to do this, they need to be prepared to transform or change in line with the changing needs of the market. In this sense, **positive organisations are not afraid of the need to change processes, reallocate resources or even operate in a different market with different products** However, it is important that all changes/reconfigurations that take place in the organisation are managed efficiently, because an excellent idea can be ruined if its implementation is poorly managed	McGrath [20] Macdonald et al. [2]
Positive organisations recognise the importance of the communication process. In this sense, they **have transparent and fluid communication between the various hierarchical levels**. These organisations clearly and transparently share information that is relevant to their success and that of their employees, such as the appraisal process, the corporate objectives and those of each employee	Rego [23]
The organisational culture of positive organisations does not tolerate behaviour that harms other employees. In this sense, **these organisations reward respect and tolerance between employees, so there is no room for bullies or corporate psychopaths** By adopting this behaviour, these organisations are investing in employee satisfaction, which has a major impact on productivity, job performance, absenteeism and staff turnover	Boddy and Taplin [10]

(continued)

Table 2.2 (continued)

Defining the concept of positive organisations	Authors
The happiness of employees in an organisation has a major impact on their productivity. In this sense, **positive organisations are concerned about the well-being of their employees, as well as the balance between their professional and personal lives.** These organisations also offer a benefits package that enables their employees to have a sustainable financial life	Nielsen et al. [24] Allen [25]
Innovation plays a key role in an organisation's competitiveness. **Positive organisations recognise the importance of innovation and ensure that the organisation reconfigures itself and implements innovations** in order to remain relevant in the market	Au-Yong-Oliveira [5, 6]

of working, manufacturing, different organisational cultures, among others, realising how these are interconnected with their organisation's competitiveness when compared to foreign organisations [26].

Globalisation increasingly represents a critical factor for an organisation's competitiveness, and this is one of the great challenges of the century for organisations, as it adds more competition to the same market. It is therefore essential for organisations to be able to implement strategies that enable them to compete with their rivals worldwide [3]. Globalisation has made it possible for countries and their organisations to be increasingly interconnected, with greater distribution of goods/products/services, and financial capital (money) and knowledge (data vs. information) [3], with the Internet being another of the driving forces behind the interconnection of humanity across countries, oceans and languages [26] and therefore organisations.

Globalisation has also led to the creation of a concept of organisations at the national level. This means that, as well as competitiveness being measured between organisations, globalisation has also allowed this competitiveness to be measured between different countries [27]. Expanding on this concept, Porter [27] defined the factors that determine a nation's Competitive Advantage, which consist of (i) Factor Conditions, (ii) Demand Conditions, (iii) Related and Supporting Industries and (iv) Firm Strategy, Structure and Rivalry.

i. **Factor Conditions** are necessary to compete in a given industry as they reflect the nation's position in terms of factors of production, such as skilled labour or infrastructure.
ii. **Demand Conditions** correspond to the nature of demand in a domestic market for the industry's product or service.
iii. **Related and Supporting Industries** correspond to internationally competitive industries or the presence/lack of supplier industries in the country.
iv. **Firm Strategy, Structure and Rivalry** are the conditions that govern how organisations are created, structured and managed, as well as the nature of domestic rivalry. Domestic rivalry favours the creation of innovative solutions.

As can be seen in Fig. 2.1, these four factors define the diamond of competitive advantage, all of which are interrelated and affect each other. Competitive advantage comes from the organisation's leadership, which shares and amplifies the forces in the diamond to promote innovation [27].

On the other hand, according to author McGrath [20], companies should not just have a competitive advantage over their competitors, they should not use frameworks and practices devised in a completely different era to today, such as Michael Porter's diamond of competitive advantage. According to the author, the premise defended by Porter [27] rewards and values stability, however, given the constant changes inherent in today's global market, stability can be a very dangerous state for a company's sustainability. When the organisation and its managers seek stability in their business, they create all the wrong reflexes, as they allow employees, regardless of their place in the hierarchy, to fall into their day-to-day routine. Managers cannot ignore possible changes in their business and therefore must be prepared to make the necessary changes, including the reallocation of resources. In order to be competitive, a company must be prepared to change whenever necessary, and so they need to detach themselves from their successful past in order to focus on the steps they need to take, the measures they need to implement and the changes they need to make in order to remain competitive—the theory of transitional competitive advantage [20].

It is also important for organisations to be able to recognise business opportunities and have the ability to reallocate their resources in order to meet their customers' needs and achieve a transitional competitive advantage. This transitory competitive advantage will allow them to remain competitive in a constantly changing market, and it is this type of competitive advantage that organisations should strive for. After all, if the organisation adapts to the changes presented to it, its service or product will also adapt, which will allow it to remain competitive.

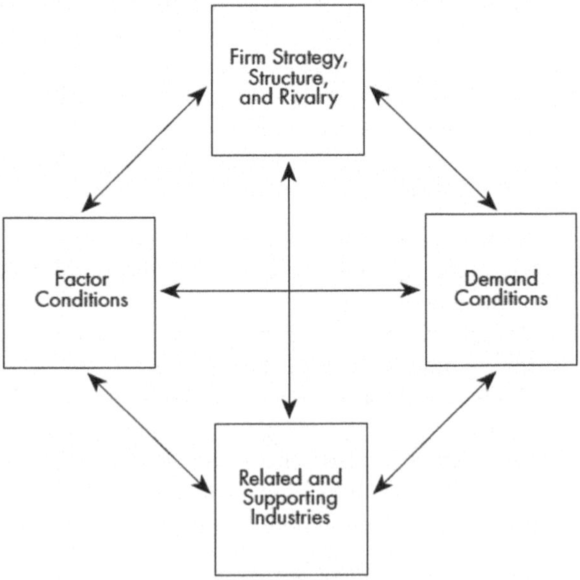

Fig. 2.1 Porter's diamond of competitive advantage. Taken from Porter [27]. The Competitive Advantages of Nations. Harvard Business Review, March–April, pp. 73–91

This is one of the guidelines of the theory of transitional competitive advantage, which argues that in order to remain competitive, an organisation must be prepared to recognise opportunities to meet market needs, regardless of the industry in which they operate [20]. As mentioned above, recognising these opportunities means having the right person for each role, performing the right function [2].

2.2.1 The Importance of Innovation

In an extremely volatile and global world, innovation, especially in recent decades, has assumed a key role in an organisation's competitiveness and survival, as it is believed that a company that does not innovate is destined to fail [5, 6]. The list of organisations that have failed to be relevant, or even disappeared, because they stopped innovating, is long. Many of these companies have stagnated and failed to remain relevant to current market needs. In this sense, however difficult it may be to implement transitions and changes, this implementation is crucial for the survival of the organisation and proves to be the right attitude when facing an environment that is constantly changing [20].

Innovation is not just about opening up new markets, as organisations can also innovate in well-established and mature markets. The key to innovation is the ability to make connections, spot opportunities and capitalise on them [28]. Furthermore, the innovation process, like any other organisational process, can be managed, which seems to make it easier to implement within the organisation. Managing this process efficiently and effectively is very important for its success, as it is this process that allows the organisation to stay ahead of its competitors and thus be more competitive. Although innovation is an organisational process that can be implemented by all organisations, it is important to note that organisations tend to make changes throughout the process and are thus able to see the suitability of certain aspects of the process for their specific organisation [20].

An organisation can adopt an open or closed innovation process, both of which have their advantages and disadvantages. In an open innovation process, the organisation collaborates with entities and/or individuals outside the organisation, such as universities or research centres. On the other hand, in a closed innovation process, the organisation only collaborates with internal individuals, and sometimes not all employees have access to the innovation process [29]. The closed innovation process used, for example, by Apple—a leading technology company—prevents copying in the market. Therefore, in this context, secrecy (both internal and external) is seen as a good way to create innovative products and services (relying solely on internal resources) [5]. Regardless of the method of innovation it adopts, it is important that the organisation implements a continuous process of innovation without interruption, because, if it exists, the lack of continuity in this process conveys the wrong idea to its employees. It is therefore necessary for the organisation to set a budget, as well as integrate this process as part of its employees' career plans [20].

In addition, it is important for the organisation to encourage and reward "innovation champions", i.e., individuals who informally emerge to promote innovations with conviction, persistence and energy, and who are also very important so that ideas do not remain "dormant" or "forgotten". These individuals voluntarily risk their position and reputation within the organisation to ensure the success of the innovation, whether self-proposed or proposed by others. Innovation champions are recognised as pivotal because they can overcome fierce organisational inertia and/or opposition and thus work to ensure that new ideas are implemented. Effective innovation champions distinguish themselves from other employees by (i) conveying confidence and enthusiasm about innovation, (ii) enlisting the support and involvement of key stakeholders and (iii) persisting in the face of adversity [12]. It is important to note that the emergence of "innovation champions", and innovation itself within the organisation, will not materialise without the organisation's efforts in this direction. This means that there must be investment within the organisation, across all departments, in R&D (Research & Development), which fosters the emergence of innovative ideas, products and services, and the prominence of these champions in relation to other employees [6]. Innovation and constant investment in R&D allow organisations to become competitive in their domestic markets as well as in global markets. However, innovating and being innovative is not easy, as company leadership is often too focused on the organisation's core activities [27].

Finally, according to Amabile [15], fostering creativity among employees also fosters innovation within the organisation, and is therefore a "powerful competitive weapon", as it allows new ideas to emerge, as well as stimulating the notion of commitment within the organisation. Being creative means loving the process of innovation [30]. Furthermore, in order to have a successful innovative company, it is necessary to bring together creative people and organise their efforts. However, it is important to recognise that innovation often occurs in an unplanned, accidental and unintentional way [31].

All of the above are tangible steps that can be taken by organisations to foster a more innovative environment, but they can also constitute major changes not only to the organisational structure, but also possibly to its core business. However, aversion to change is the main enemy of innovation, as organisations and their employees have the feeling that they "have a lot to lose" by implementing new processes or investing in an innovative product. The internal environment of organisations therefore functions as an "immune system" that rejects change, causing the organisation to stagnate and consequently lose its competitive advantage, which translates into its competitors overtaking it [27].

In short, companies must investigate and develop new ways of becoming more innovative. Thus, only by continuously innovating will companies succeed in the market and distinguish themselves from their competitors. Innovation is not only important for the organisation itself, but also for national economic growth and, at a later stage, for global economic growth, since, if successful, it can influence the global market [28].

Finally, as explained above, there are concrete steps that organisations can take to foster their innovative streak, such as (i) fostering creativity, (ii) recognising innovative employees who contribute to the organisation's innovation and (iii) strategically defining where the organisation wants to position itself and, consequently, understanding how investment in R&D and innovation can help it achieve these goals.

2.2.2 *The Role of Leadership*

The leaders of organisations play an essential role in the smooth running of the organisation and are therefore largely responsible for its success or failure [32]. The term "Leadership" has been widely explored in the literature [2, 32]. According to the existing literature, a leader is an individual who is able to demonstrate their power or authority in order to influence those under them to act in a certain way so that the organisation achieves its goal. For an organisation to function efficiently, it is important that all leaders are clearly identified and defined within the organisation, thus facilitating the execution of work effectively and efficiently [2].

The role of a leader is a social process, in which the leader influences the group, without the use of force, manipulation and/or power, in order to achieve the objectives he or she proposes, which are aligned with the overall objectives of the organisation. In addition, a good leader always has the well-being and success of their organisation at heart, so they must initially observe and listen to it and then act on what they have learned, as long as they do so with the organisation's interests in mind. Your popularity and the reaction of employees to the decisions you make cannot be a factor in influencing the measures you implement. This means that he or she has to make the decisions that will be best for the organisation, such as reducing costs or introducing changes to the work process, without his or her decision being influenced by others [2]. Thus, a leader's job is to create, maintain and improve the unique and productive culture of a group of people, so that they achieve the objectives outlined and continue to do so over time.

In addition, it is the leader's role to show the company's employees that it is changeable to the needs and changes of the market, as discussed in Sect. 2.2, and that it is effectively in constant transition with these changes in order to remain competitive in the market in which it operates. Organisations therefore need leaders with a *mindset* that knows how to identify future business opportunities and act accordingly, as well as the ability and willingness to foster change, the ability to resolve possible setbacks and conflicts and, consequently, devise appropriate responses to the various problems the organisation may face [20]. As well as having a direct influence on the path the organisation takes, the leadership of an organisation also affects the well-being of employees. In a negative organisation, the leadership allows and encourages the use of lies and manipulation as a means of career progression, so competent, honest and loyal employees lose their place and end up leaving [5]. This can mean a loss of knowledge, innovation and competitiveness.

A practical example of the influence of leadership on the organisation is the case of Kodak and Netflix. Unlike Kodak, which insisted on focusing on its core product—roll cameras—with no room for innovation, the leaders of the Netflix company recognised that all organisations, regardless of their size, have to adapt and reconfigure themselves if they are to thrive [20, 33]. Netflix, initially a DVD rental company, has recognised how to do business to keep up with trends, pioneering digital transformation in the entertainment distribution market [33]. Kodak, on the other hand, believing that its product would resist market changes and innovations, chose not to introduce innovative products, which led to a loss of competitiveness and market share.

It is therefore important for leaders to have certain characteristics and behaviours that allow them to stimulate the growth of the organisation and promote the development of an innovative and competitive mindset in all employees. These characteristics enable them to understand the changes and uncertainties of the future and to bring together characteristics that allow them to better predict it, as well as how they can lead their organisation more efficiently, taking this prediction into account [34]. According to Pistrui [34], these characteristics are:

1. **Increasing sensitivity**: Managers need a change of mentality and a commitment to the organisation in order to increase their sensitivity to the changes in the market in which they operate, and thus make strategic decisions for the growth of the organisation. An example of this phenomenon is Uber—while for some, this company is just an "app for hailing taxis", for others, Uber represents a new platform for taxi services. In addition, there are those who see Uber as a natural step towards a future in which taxis will be autonomous. This means that some leaders are already able to foresee, by having broadened their awareness, that the taxi service of the future could be very different from the current one, which allows them to make decisions and implement measures in line with this prediction.
2. **Defending change**: When a leader commits to a new direction, it is crucial that there is no doubt about their commitment to following that path. Although uncertainty is intrinsic to all change processes, leaders must outline the new path the organisation will take, as well as how this new path will alter the organisation's priorities. According to Pistrui [34], the companies most likely to survive today are those with leaders who personally defend change and promote it.
3. **Creating a new order**: Championing change implies creating a new agenda and adopting a new point of view, which requires restructuring the company and triggering the actions needed to achieve goals that may never have been defined or reached. The creation of a new order begins when leaders establish new priorities and norms that are more productive and in harmony with the new path the company has to take [34].
4. **Leading with a sense of anticipation**: When a presumed assumption becomes a new corporate vision, it is necessary to lead the entire company in this direction. Employees who are tied to *status quo* forms of success must be motivated to adopt different behaviours. Therefore, leaders must remove all doubt that

everyone needs to develop the skills and behaviours that will enable them to achieve something new and extraordinary.

Each leader should therefore seek to develop the four competences mentioned above, adapting them to their organisation and shaping them into their own leadership approach. If the organisation's leadership is uncertain about its future, this uncertainty can have negative consequences for the organisation, generating widespread confusion about the direction the organisation needs to take and, consequently, little enthusiasm among the company's employees. In this way, cultivating these four competences will enable the leadership circle to clarify the next step for the organisation and direct its management approach towards the future [34].

It is, however, important to clarify that managers have no way of predicting the future with absolute certainty, especially with regard to (i) the evolution of an industry, (ii) customers and their preferences, (iii) the organisation and its potential capabilities, as well as its cost structure and (iv) its competitors [35]. Even so, despite the uncertainty and complexity of leading a company in a global, competitive and innovative market, leaders cannot forgo making strategic choices that are in line with their vision of what the organisation could be Martin [35]. Therefore, managers are required to have a clear vision of the organisation's strategy and future objectives [36] even if they are not in line with current objectives.

According to Webb [36], futurists have a different perspective about time and as such organisational strategists have much to learn from their approach, i.e., as a rule, when organisations are faced with future uncertainty—for example, a risk or an opportunity—they tend to think in sequential timelines. According to the author, this is not the best approach, as it is necessary to think about the organisation's short-term and long-term objectives simultaneously. To this end, it is recommended that organisations apply a "Futurist's Framework for Strategic Planning".

This conical model (see Fig. 2.2) means that as the organisation progresses, the cone will also adapt, taking into account the organisation's progress in meeting the objectives outlined previously, as well as external factors unknown to the organisation at the time the model was initially defined. In this way, the model gives the organisation flexibility to respond to unexpected situations. As can be seen in Fig. 2.2, this model is made up of four distinct categories: (i) tactics—trends and likely events, whether internal or external to the organisation; (ii) strategy—traditional strategy and the organisation's likely trajectory; (iii) vision—formulating a plan to define how the organisation is going to achieve its objectives, as well as how it is going to be able to attract skilled employees and/or train current employees, thus defining areas for future investment; and (iv) systems evolution—what concerns the evolution of the industry and the challenges that the evolution of technology brings to the organisation. It is important to note that as the organisation progresses through the categories of the established framework, the degree of uncertainty and concrete data about its success decreases. In this sense, as the organisation obtains data and evidence of the success of implementing this framework, a reassessment of what the company's future will be should be made, which translates into restarting the cone

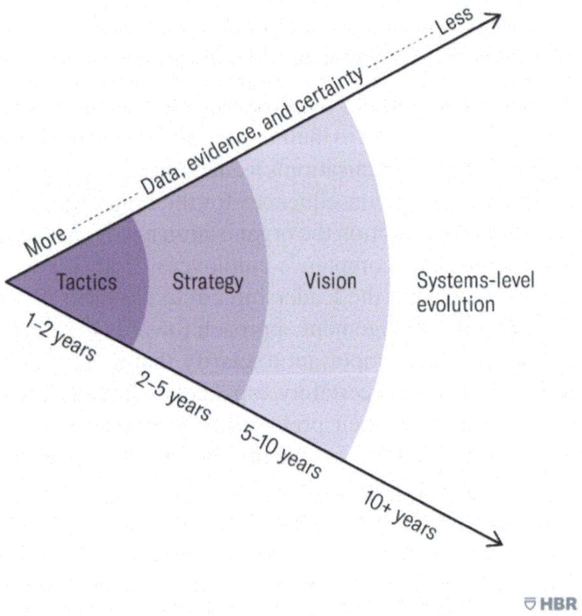

Fig. 2.2 Futurist's framework for strategic planning [36]

and its tactical category in the present. The result, ideally, is a flexible organisation that is positioned to continuously interact and respond to external developments [36].

In short, a good leader must have the right characteristics to lead the organisation towards a successful future, which means that he or she must maintain a mindset that is firm in its support, flexible to internal or external changes, and futuristic in the implementation of the necessary measures and changes in the organisation's direction.

2.2.3 The Role of Government

As has been mentioned in previous chapters, a country's competitiveness depends essentially on the ability of organisations to innovate and change whenever necessary, so it is clear that national prosperity is not hereditary, but created [27]. As you might expect, the government also has a role to play in this process. The role of governments should be to challenge and encourage organisations to be more competitive and to reach their full potential, not just to survive, but to be competitive in the international marketplace [27]. Governments are not responsible for creating competitive industries, but they can implement measures and regulations that favour innovation and, consequently, competitiveness. However, as innovation is a prolonged process, governments usually favour short-term measures and are not concerned with the nation's competitive future, thus delaying innovation and its potential [27].

In addition, many organisations share the myopic vision of their governments, and only give importance to what has a direct impact on their day-to-day activities, not paying attention to trends and events taking place in another part of the globe [26]. However, in an increasingly global world, there are trends and events whose impact goes far beyond the place and organisation in which they occur. These are called megatrends and can affect governments, economies, societies and cultures [26]. According to Tse and Esposito [26], megatrends have a universal impact, regardless of their location or economic situation, as they represent disruptive patterns or changes for organisations as well as society as a whole. Their study revealed the existence of five megatrends—(i) Demographic and Social Changes, (ii) Scarcity of Resources, (iii) Inequalities, (iv) Volatility, Scale and Complexity and (v) Entrepreneurial Dynamics—which help individuals, organisations and governments to identify market opportunities in attractive clusters, as well as providing guidance on how the future will unfold, which implies a joint and transversal analysis of all these factors [26]. Table 2.3 explains each of the trends in greater depth.

In view of the trends and challenges of the future, a profound change in mindset—on the part of companies, the government and the general public—is essential. Only in this way will they be able to meet the challenges that the trends of the future bring. It is therefore necessary for all three players in the process to think of the future as engineers and not as mere observers, thus anticipating possible needs and problems. Governments in particular should help organisations identify market opportunities according to the most attractive clusters so that they invest in them.

It is therefore up to governments to implement measures that encourage continuous innovation and its implementation in the day-to-day running of organisations. These measures can come from the most varied hierarchies of the country's government, but they must serve the interests of the country as a whole and, in particular, the interests of the organisations in which they will be implemented. As regards Portugal in particular, as it is a country that does not appreciate change, the Portuguese have a preference for staying in a bad situation rather than accepting a change for the better [37]. According to the European Innovation Scoreboard [38], Portugal is a moderately innovative country [38]. In addition, negative organisations and their less positive working environments are very prominent in countries like Portugal—relationship oriented and averse to change [4, 37]. Company leaders and governments, even though they want to implement measures that bring about significant improvements, are unable to implement solutions that make a difference due to the existing *status quo*.

In conclusion, it is important for organisations, the government and the general public to realise that the better we deal with current problems, the better the chances of creating a better future. If we do not act today, tomorrow will not bring improvements [26]. Analysing the aforementioned trends is therefore essential for companies to be "ahead of the curve", i.e., to be able to identify possible future risks and opportunities and thus have a strategic mindset for the future.

Table 2.3 Megatrends—(i) Demographic and social change, (ii) Scarcity of resources, (iii) Inequalities, (iv) Volatility, scale and complexity and (v) Entrepreneurial dynamics [26]

D	Demographic and social change
	People in general are at the centre of the discussion about the future in terms of how it affects organisations. This trend is marked by (i) the ageing of the global population, with fewer and fewer active contributors, (ii) an increase in the retirement age to try to compensate for the evolution of the last point and (iii) the growing migration of the population to the cities, where there are more job opportunities
R	Scarcity of resources
	The use of resources is very important for our society, because it is through them that society grows and advances. Currently, water and food are the main resources that are under threat of scarcity, mainly due to (i) the growth of cities and urbanisation and, consequently, the increase in the rate of consumption, (ii) climate change and (iii) food waste. Given the indispensability of these resources for human life, it is important that individuals, organisations and governments think about how they can consume fewer resources, i.e., implement a circular economy in their daily lives so that they need to use fewer resources by reusing existing ones
I	Inequalities
	The distribution of resources has been unequal on different continents for centuries. Inequalities have to do with (i) monetary differences, (ii) access to resources, such as education, (iii) age differences, since older generations have access to more advantages and luxuries than millennials or generation Z will ever have, (iv) access to politics and politicians, since wealthier individuals and organisations have the ability to influence a country's policies
V	Volatility, scale and complexity
	The world is currently seeing a great deal of volatility due to rapid changes in technological development, from artificial intelligence to robots automating jobs previously done by humans. In the past, volatility was seen as a problem and so the main objective was to avoid change. The new norm is to accept that the world is changing, and will continue to change, at a rapid pace, so volatility should be seen as an opportunity Even though machines and robots are replacing individuals in their work, new job opportunities will arise naturally, so the important thing is for individuals to continually invest in their training in order to be relevant to the organisation. This change also allows organisations to be increasingly productive
E	Entrepreneurial dynamics
	Major changes are taking place in the way work is carried out, as well as the way commerce is operated. The speed of change has allowed new business models to emerge, requiring existing companies, regardless of their size or length of existence, to be "forced" to constantly think about their next steps, which is why innovation plays a key role in today's and tomorrow's business society. Innovation brings about changes in practice, approach and mindset

References

1. Stewart, D.W.: Focus Groups. The SAGE Encyclopedia of Educational Research, Measurement, and Evaluation, vol. 2, pp. 687–692. Sage Publications, Thousand Oaks (2018)
2. Macdonald, I., Burke, C., Stewart, K.: Systems Leadership: Creating Positive Organisations, 2nd edn. Routledge Publishing (2018)

3. Hitt, M., Ireland, R., Hoskisson, R.: Strategic Management: Concepts. Competitiveness and Globalisation, 9th edn. South-Western, Mason, USA (2011)
4. Solomon, C.M., Schell, M.S.: Managing Across Cultures–The Seven Keys to Doing Business with a Global Mindset. McGraw-Hill, USA (2009)
5. Au-Yong-Oliveira, M.: Negative Organisations and How to Overcome Them. Fighting to Promote Innovation and Change. Sílabas & Desafios, Faro (2022)
6. Au-Yong-Oliveira, M.: Negative organisations and (negative) powerful relationships and how they work against innovation-perspectives from millennials, generation Z and other experts. Sustainability **14**, 17018 (2022). https://doi.org/10.3390/su142417018
7. Fernandes, P.: Corruption: Definition, Prevention and Combat-The Case of S. Tomé and Príncipe. University of Aveiro. Autonomous Section of Social, Legal and Political Sciences. Aveiro (2009)
8. Hofstede, G., Hofstede, G. J., Minkov, M.: Cultures and Organisations: Software of the Mind. McGraw-Hill (2010)
9. Au-Yong-Oliveira, M., Lebre, I.A.P.M., Nogueira, A.R., Gonçalves, R.: Success, failure, marketing and innovation: the case of Nokia. RISTI-Iber.N J. Inf. Syst. Technol. 219–234 (2020)
10. Boddy, C.R., Taplin, R.: The influence of corporate psychopaths on job satisfaction and its determinants. Int. J. Manpow. 965–988 (2016)
11. Moore, J.F.: Predators and prey: a new ecology of competition. Harv. Bus. Rev. **71**(3), 75–86 (1993)
12. Howell, J.M.: The right stuff: identifying and developing effective champions of innovation. Acad. Manag. Exec. **19**(2), 108–119 (2005)
13. Taylor, B.: Don't let negativity sink your organisation. Harvard business review (cited February 2020). Available from: https://hbr.org/2020/02/dont-let-negativity-sink-your-organization on 15 July 2022
14. Vickers, M.H.: Towards reducing the harm: workplace bullying as workplace corruption-a critical review. Empl. Responsib. Rights J. **26**, 95–113 (2014)
15. Amabile, T.M.: How to kill creativity. Harvard Business Review (cited September-October Issue). Available from: https://hbr.org/1998/09/how-to-kill-creativity (1998)
16. Schein, E.H.: Organisational Culture and Leadership, 3rd edn. Jossey-Bass (2004)
17. Oliveira, M.: Building leadership: how pride in your work leads to better attendance and quality. Perform. Improv. **44**(7), 5–7 (2005)
18. Tupper, H., Ellis, S. It's Time to Reimagine Employee Retention. Harvard Business Review (2022)
19. Boddy, C.R.: Corporate psychopaths, bullying and unfair supervision in the workplace. J. Bus. Ethics **100**, 367–379 (2010)
20. McGrath, R.G.: The end of competitive advantages: how to keep your strategy to move as fast as your business. Harvard Business Review (cited August 2013). Available from: https://hbr.org/2013/08/the-end-of-competitive-advanta (2013)
21. it depends upon us: Tjosvold, D. (2007). The conflict-positive organisation. Journal of Organizational Behaviour **29**, 19–28 (2008)
22. Tjosvold, D., Su, F.: Manager as mediator: devoloping a conflict-positive organisation. In: Herrman, M.S. (ed.) The Blackwell Handbook of Mediation: Bridging Theory, Research, and Practice. Wiley & Sons (2006)
23. Rego, A.: Personal and Organisational Communication: Theory and Practice, 1st edn. Edições Sílabo, Lda (2007)
24. Nielsen, K., Randall, R., Yarker, J., Brenner, S.O.: The effects of transformational leadership on followers' perceived work characteristics and psychological well-being: a longitudinal study. Work Stress. **22**(1), 16–32 (2008). https://doi.org/10.1080/02678370801979430
25. Allen, T.: The pandemic is changing employee benefits. Harvard Business Review. (cited April 2021). Available from: https://hbr.org/2021/04/the-pandemic-is-changing-employee-benefits (2021)

26. Tse, T.C., Esposito, M.: Understanding How the Future Unfolds: Using DRIVE to Harness the Power of Today's Megatrends. Lioncrest (2017)
27. Porter, M.: The Competitive Advantages of Nations, pp. 73–91. Harvard Business Review, March-April (1990)
28. Tidd, J., Bessant, J.: Managing Innovation: Integrating Technological, Market and Organisational Change, 7th edn. Wiley (2013)
29. Enkel, E., Gassmann, O., Chesbrough, H.:. Open R&D and open innovation: exploring the phenomenon. Blackwell Publishing Ltd. Available from: https://onlinelibrary.wiley.com/doi/abs/, https://doi.org/10.1111/j.1467-9310.2009.00570 (2009)
30. Amabile, T.M.: Motivating creativity in organisations: on doing what you love and loving what you do. Calif. Manage. Rev. **40**(1), 39–58 (1997)
31. Rosenberg, N.: Why do firms do basic research (with their own money?). Res. Policy **19**(2), 165–174 (1990)
32. Mitchell, T.R., Larson, J.R., Jr.: People in Organisations: An Introduction to Organizational Behaviour, 3rd edn. McGraw-Hill (1987)
33. Au-Yong-Oliveira, M., Marinheiro, M., Tavares, J.: The power of digitalisation: the Netflix story. WorldCIST. **2**, 590–599 (2020)
34. Pistrui, J.: How managers can see the future more clearly. Harvard Business Review (cited October 2015). Available from: https://hbr.org/2015/10/how-managers-can-see-the-future-more-clearly (2015)
35. Martin, R.L.:. Strategy is iteractive prototyping. Harvard Business Review (cited June 2014). Available from: https://hbr.org/2014/06/strategy-is-iterative-prototyping (2014)
36. Webb, A.: How to do strategic planning like a futurist. Harvard Business Review. Available from https://hbr.org/2019/07/how-to-do-strategic-planning-like-a-futurist (2019)
37. Hofstede Insights Oy.: Hofstede insights-COUNTRY COMPARISON TOOL. Available from: https://www.hofstede-insights.com/country-comparison-tool?countries=portugal in September 2023
38. European Commission.: European innovation scoreboard 2021-coutry profile Portugal. Available from: https://ec.europa.eu/assets/rtd/eis/2023/ec_rtd_eis-country-profile-pt.pdf (2021)

Chapter 3
Methodology

Abstract The research methodology followed for a research project needs to be communicated in as transparent a fashion as possible, so that other researchers may understand how the study was performed and how the data was collected. If other researchers intend to replicate the study, in another context, this should also be possible after reading the methodology section or chapter. The research question is hence focused on in this section, as is the PRISMA methodology (systematic literature review) and a description of how the interviews and focus group were undertaken (qualitative data collection and research effort). The methodology section is crucial in a research project and often is what separates academic research from journalism. Experienced researchers will often first turn to the methodology section to ascertain what method was followed as if the method is not sound then the rest of the study may not be worth reading, as it will have an erroneous basis.

Research can be defined as a process which, with a clear objective, is carried out in a systematic way so that, as a result of the research performed, new discoveries can be made. A systematically designed investigation involves an explanation of the method used to collect the data, a pertinent justification for the relevance of the data collected, as well as the limitations associated with the research method. In this sense, we can see that research is characterised by (i) clearly delineated objectives in order to discover new realities, (ii) systematically collected data and (iii) systematically interpreted data [1].

This chapter aims to transparently convey all the procedures used to carry out this research, as well as the relevance of the choice of topic.

3.1 Formulating the Research Question

Consumers are giving organisations less and less opportunity to make mistakes and to follow less ethical choices. It is therefore imperative that organisations change the way they formulate their strategies, not only to take this into account, but also to

B. Graça and M. Au-Yong-Oliveira, *Competitiveness Strategies for Negative Organizations*, Information Systems Engineering and Management 1,
https://doi.org/10.1007/978-3-031-59914-9_3

remain competitive in a fierce, global marketplace [2]. It is therefore expected that all the strategies outlined in companies will not only serve the purpose of making them more competitive, but even more "positive". However, despite the authors' combined numerous years of professional experience, they have noted that the concept of a "positive organisation" is often an illusion, even though in the academic and professional world it is "sold" that organisations are a perfect place to work.

This research—Competitiveness Strategies for Negative Organisations: Challenging the *Status Quo*—aims to address the subject of negative organisations, with the aim of providing organisations, government and the general public with strategies to make their workplace a more positive and competitive environment. The main aim of management research is to help organisations achieve sustainable growth in a successful way, thereby promoting the development of innovation [3]. **The research question of this investigation relates to identifying the problems that influence the competitiveness of negative organisations**. Using the information gathered in this study, the aim is to develop strategies that will enable Portuguese organisations to overcome the *status quo* and the "successful" past in order to become and/or remain competitive in the future. As stated above, given that innovation is an inherent aspect of positive and generally competitive organisations, the creation of a methodology that allows stagnant organisations to get back on track and become competitive is extremely important.

3.2 Research Method

As mentioned above, this research aims to provide Portuguese negative organisations with a set of strategies so that they can become positive organisations in the future. This research adopts a qualitative methodology, i.e., an analysis based on qualitative data [4], mainly words. This type of analysis involves the data being interpreted by the researcher, as a description of a hypothesis, theory or case study is developed, an investigation is carried out that culminates in the generation of qualitative data and, subsequently, the researcher analyses this data [5].

In methodological terms, this study consisted of three main phases, carried out consecutively:

i. Preparation of the literature review, in which the topics relevant to the analysis were studied in order to gain a clear understanding of what a negative organisation is and how they damage their own competitiveness. To this end, this research followed a systematic literature review on the SCOPUS platform—see Tables 3.1, 3.2 and 3.3—as well as a non-systematic review of various sources relevant to this study, namely the *Harvard Business Review*. In addition, a PRISMA methodology was also developed. These approaches made it possible to identify gaps in existing knowledge, especially with regards to negative organisations and their impact on competitiveness, as well as to understand the reliability and applicability of the results gathered.

Table 3.1 Search on the Scopus platform for "Negative Organisations"

Contents	Explanation
Research objective	To see whether the topic of Negative Organisations has been investigated by a significant number of researchers, and to see how recent the interest in it is
Database	Scopus, as it is an academic database that is widely recognised and used by researchers in various fields of knowledge, due to its comprehensiveness, quality of content, advanced search resources, citation metrics and continuous updating
Research date	11th December 2021
Selected criteria	All sources; with "Negative Organisations" in the title, abstract and keywords; All areas of research; All years
Total number of articles found	9 Articles
Date of oldest article	2002
No. of articles relevant to this study	4 Articles

Table 3.2 Research on the Scopus platform on "Corporate Psychopaths"

Contents	Explanation
Research objective	To understand whether the topic—Corporate Psychopaths—has been investigated by a significant number of researchers, as well as how recent the interest in it is
Database	Scopus, as it is an academic database that is widely recognised and used by researchers in various fields of knowledge, due to its comprehensiveness, quality of content, advanced search resources, citation metrics and continuous updating
Research date	11th December 2021
Selected criteria	All sources; with "Corporate Psychopaths" in the title, abstract and keywords; All research areas; All years
Total number of articles found	43 Articles
Date of oldest article	2010
No. of articles relevant to this study	19 Articles

Table 3.3 Research on the Scopus platform on "Bullying and Corporate Psychopaths"

Contents	Explanation
Research objective	To understand whether the topic—Bullying and Corporate Psychopaths—has been investigated by a significant number of researchers, as well as how recent the interest in it is
Database	Scopus, as it is an academic database that is widely recognised and used by researchers in various fields of knowledge, due to its comprehensiveness, quality of content, advanced search resources, citation metrics and continuous updating
Research date	11th December 2021
Selected criteria	All sources; with "Bullying" and "Corporate Psychopaths" in the title, abstract and keywords; All research areas; All years
Total number of articles found	7 Articles
Date of oldest article	2011
No. of articles relevant to this study	7 Articles

ii. Preparing and carrying out six individual interviews, as well as one focus group with managers and employees of organisations, with the aim of gathering different and/or complementary perspectives on this study. As well as gathering relevant information, this stage aims to promote change within the organisations of which the participants are a part because, assuming that there are no totally positive organisations, there is always room for improvement.
iii. Analysis of the data and perspectives gathered and the creation and formulation of a strategy that will enable Portugal and its organisations to be more prosperous and competitive in the future (see Fig. 3.1). **It should also be noted that although this project is a study of organisations, it is considered that the role of government entities is essential for Portuguese organisations to achieve their purpose.**

3.2.1 PRISMA Methodology

As mentioned above, the PRISMA methodology—The Preferred Reporting Items for Systematic reviews and Meta-Analyses—has the main objective of verifying the reliability and applicability of the results of the literature review, so the development of this methodology is intended to deliver the results in a transparent and complete manner [6].

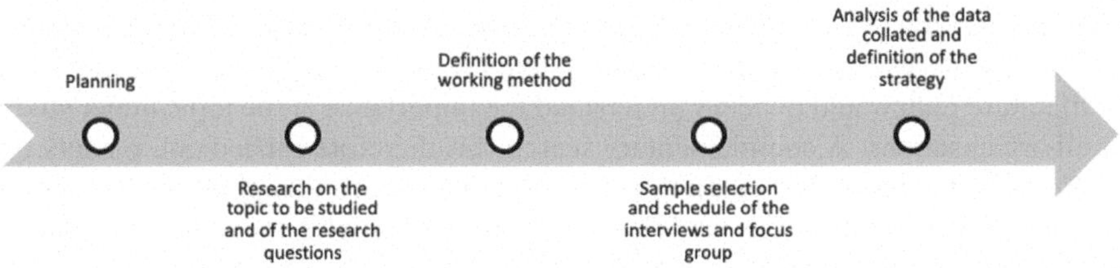

Fig. 3.1 Planning the research method and all its phases

According to the information mentioned above, the aim of this research is to explore and understand the concept of a negative organisation, as well as to analyse how such organisations can damage their own competitiveness. This research therefore aims to identify and analyse the main factors that contribute to the creation of negative organisations and how these factors impact on their competitiveness. In addition, it will examine the possible strategies and measures that organisations can adopt to reverse this negative trend and, consequently, improve their competitive position. The aim of this literature review is to contribute to a deeper understanding of the implications of a negative organisation and the actions needed to promote a healthy and sustainable organisational culture.

The "Prisma 2020 Checklist" (see Appendices 1 and 2) was used when carrying out this methodology, in order to identify all the steps and information to be included in this study, as well as to realise whether this analysis would be complete and adequate for it.

It is important to clarify that, when this research began, the state of knowledge was very limited and there was not much knowledge developed in the main area under study—Negative Organisations. Even so, the researchers felt that studying this topic would be imperative for the field of management. As mentioned above, a search was carried out on the SCOPUS platform, the results of which are shown in Tables 3.1, 3.2 and 3.3.

It is also important to note that the selection of relevant articles for the study followed the following steps:

1. Search on the SCOPUS platform, according to the criteria mentioned in Tables 3.1, 3.2 and 3.3.
2. Preparation of an Excel document for each of the studies, which contained all the information relevant to the study: (i) authors, (ii) title and year, (iii) DOI, (iv) link to the document, (v) abstract and (vi) keywords.
3. Analysing the information that each of the Excel documents contained and therefore, through this analysis, checking which articles were of potential interest to this study. The number of articles selected can be seen in Tables 3.1, 3.2 and 3.3 in the line "No. of articles relevant to this study".
4. Printing all the articles considered relevant, analysing them and then integrating them into this study.

As can be seen from Tables 3.1, 3.2 and 3.3, the knowledge available was limited, so it was also necessary to resort to a non-systematic methodology in order to carry out the literature review and to better understand the importance of the topic under study for all organisations. A complementary search was therefore carried out, namely on the *Harvard Business Review* platform. The authors chose this platform because it has great global credibility, as well as a wide variety of articles and topics that helped to complement this research.

In this sense, you can find the flow diagram in Appendix 3, which explains the sources of all the bibliography analysed and how many studies were actually used.

3.3 Participant Selection and Interviews

As mentioned above, this study, in addition to the literature review carried out and explained above, will include individual interviews with various employees and managers of organisations. The use of this research methodology—i.e., an intentional conversation between two or more people, during which the interviewer asks concise, unambiguous questions and listens attentively to the interviewee, in order to explore points of interest and clarify and confirm meanings—is an important part of the study [1]—allowing the researcher to get to know new points of view on the subject under study and thus develop more critical thinking that is closer to the current reality of the subject in question.

The interviews conducted followed a semi-structured methodology, which is characterised by combining predetermined questions with the flexibility to explore additional topics and allow for more elaborate answers from the interviewee. In this sense, semi-structured interviews begin with a predetermined list of topics and possibly a few key questions related to the topics being analysed, in order to guide the conduct of each interview [1]. The interviews began with a brief explanation of the topic under study in this publication, followed by a brief introduction of the interviewee. As for the pre-defined questions, these varied and were defined by the interviewer according to the interviewee's professional experience. Nonetheless, in order to make comparisons between the different perspectives, there were control questions in all the interviews, which were the same for all the interviewees.

As this is a qualitative study, the sample used plays a key role in the success of this research. This sample, provided it is carefully selected (a purposive sample), will allow us to draw relevant and pertinent conclusions [1]. In order to obtain more reliable results, the heterogeneity of the sample was also taken into account, in order to fit in with the different perspectives, so it is considered that the sample should be characterised as non-probabilistic, i.e., a type of sampling in which random forms of sample selection are not used, so sampling can be done intentionally, i.e., a purposive and heterogeneous sample [1]. The use of a purposive sample implies that the researchers use their judgement to select the cases that best suit their study, in order to meet the research objectives. In this sense, it is essential that the researchers carefully consider the impact of each of the elements used in the study. This type of

Table 3.4 Identification of selected individuals

Name	Position
João Ranito	Former senior leadership business executive and current entrepreneur
Tommy York	Software Engineer—former Google employee
Issam Ghazzawi	University Professor—University of La Verne
Veronica Perez	Assistant Marketing Director
Frank Fialho	University Professor—University of La Verne
Anonymous participant	Senior partner of a large Portuguese law firm

sample is also frequently used in studies with a small sample size [1] as is the case in this study.

In order to obtain variation in the perspectives of the sample, it is intended that the individuals are heterogeneous and have different academic and professional backgrounds [1]. Based on the aforementioned concept, this study used a sample of six individuals who, due to their professional and personal experience, would bring a relevant perspective to this study. In addition to this criterion, individuals known to the author of this study were also selected.

Table 3.4 shows the individuals selected.

In addition to the interviews conducted, a Focus Group was also held. This methodology can be defined as an informal discussion between selected individuals on specific topics [7].

The focus group participants chose to keep their contribution to this study anonymous, but for sampling purposes it is possible to access relevant information regarding their identification in Table 4.20.

3.4 Viability and Reliability

The quality and credibility of a qualitative study essentially depends on how the data is collected, so it is important for the researcher to establish guidelines such as [8]:

1. **Objectivity/Confirmability**: the researcher should conduct the research in a neutral and bias-free manner, while warning and explaining the inevitable biases that may exist. The researcher can ask the following questions, for example: (i) Are the general methods and procedures of the study described explicitly and in detail, i.e., is it clear and complete how the research was conducted?; (ii) Is the sequence of the research, i.e., how the data was collected, processed and transformed, clearly described?; (iii) Is the data obtained accessible and available so that other researchers can re-analyse it?
2. **Reliability/Dependability/Auditability**: the research should be carried out in a stable manner over time and between researchers and methods, so that there is greater control over the quality of the study in question. The researcher can

ask the following questions, for example: (i) Is the research question clearly described?; (ii) Have quality controls been carried out on the data, for example to determine possible bias?; (iii) Has the study been peer or colleague reviewed?

3. **Validity/Credibility/Internal Authenticity**: it is important for the researcher to check that the conclusions of the study make sense and are credible to the readers of the research. The researcher can ask the following questions, for example: (i) Did the triangulation between complementary methods lead to convergent conclusions?; (ii) Are the conclusions internally coherent (Eisner [9], cited by Miles and Huberman [8]) and are the concepts systematically related (Strauss and Corbin [10], cited by Miles and Huberman [8])?; (iii) Were the predictions made at the start of the research credible?
4. **External Validity/Transferability/Appropriateness**: It is important to understand whether the conclusions of this study can be transferred to other contexts and whether they can even be generalised. The researcher can ask the following questions, for example: (i) Are the characteristics of the original sample, scenarios and processes fully described to allow for adequate comparisons with other samples?; (ii) Is the sample theoretically diverse enough to encourage wider applicability?; (iii) Does the study suggest scenarios where the results could be more fruitfully tested?
5. **Use/Application/Action Orientation**: As well as it being important to understand whether the results of the research are valid and transferable to other realities, it is also crucial that the study is of practical use to the researcher, participants and readers. The researcher can ask the following questions, for example: (i) Do the research findings stimulate the reader to take specific future actions? (Lincoln [11], cited by Miles and Huberman [8]); (ii) Do the actions taken really help to solve the problem?; (iii) Have the users of the results learnt or developed new skills?

Table 3.5 shows how the aforementioned criteria are extrapolated to the research in question.

Table 3.5 Viability and reliability criteria applied to this research [8]

Objectivity and confirmability	It is important to ensure that this study was carried out impartially, following a data collection method. The data, as mentioned above, was collected through six individual interviews and one focus group, and during all the conversations there were no preconceptions that could influence the participants We can therefore conclude that it was possible to have an unprejudiced conversation with the various participants and to collect objective data for this research
Reliability, dependability and auditability	In order to confirm the reliability of this study, the research team/authors were present in all the conversations. In this way, it was possible to share the ideas that emerged from them, thus avoiding bias in this study In addition, the research question—identification of the problems that influence the competitiveness of negative organisations—was also clearly established and, in this way, it was possible to conduct this research in such a way as to guarantee its quality
Validity, credibility and internal authenticity	In addition to an exhaustive literature review, which resulted in the definition of a Negative Organisation—something under researched in the literature—two different data collection methods were also used: interviews and a focus group. The use of different complementary methods allowed the researchers to verify the credibility of the data collected
External validity, transferability and suitability	This research used a heterogeneous sample in order to collect data that was comprehensive enough to encourage holistic applicability, thus allowing comparisons to be made with different participants
Guidance on use, application and action	The aim of this research is not only to investigate the topic, but also to create tangible strategies and measures applicable to the business world, so that organisations become more positive, innovative and competitive

References

1. Saunders, M., Lewis, P., Thornill, A.: Research Methods for Business Students, 8th edn. Pearson Education Limited (2019)
2. McGrath R.G.: The end of competitive advantages: how to keep your strategy to move as fast as your business. Harvard Business Review (cited August 2013). Available from https://hbr.org/2013/08/the-end-of-competitive-advanta (2013)
3. Christensen, C.M., Raynor, M.E.: The Innovator's Solution: Creating and Sustaining Successful Growth. Harvard Business School Press, Boston (2003)
4. Elo, S., Kääriäinen, M., Kanste, O., Pölkki, T., Utriainen, K., Kyngäs, H.: Qualitative content analysis: a focus on trustworthiness. SAGE Open **4**(1), 1–10 (2014)
5. Creswell, J.W.: Research Design: Qualitative, Quantitative and Mixed Methods, 2nd edn. ARTMED Editora S.A. (2007)
6. Page, M.J., Moher, D., Bossuyt, P.M., Boutron, I., Hoffmann, T.C., Mulrow, C.D., Shamseer, L.M., Tetzlaff, J.M., Akl, E.A., Brennan, S.E., Chou, R., Glanville, J., Grimshaw, J.M.,

Hróbjartsson, A., Lalu, M.M., Li, T., Loder, E.W., Mayo-Wilson, E., McDonald, S., McGuinness, L.A., Stewart, L.A., Thomas, J., Tricco, A.C., Welch, V.A., Whiting, P., McKenzie, J.E.: PRISMA 2020 explanation and elaboration: updated guidance and exemplars for reporting systematic reviews. Research Methods and Reporting (2021)

7. Beck, L., Trombetta, W., Share, S.: Using focus group sessions before decisions are made. Med. J. **47**(2), 73–74 (1986)
8. Miles, M.B., Huberman, A.M.:. An Expanded Sourcebook: Qualitative Data Analysis, 2nd edn. SAGE Publications, Inc. (1994)
9. Eisner, E.W.: The Enlightened Eye: Qualitative Inquiry and the Enhancement of Educational Practice. Macmillan, New York (1991)
10. Strauss, A.L., Corbin, J.: Basics of Qualitative Research: Grounded Theory Procedures and Techniques. Sage, Newbury Park, CA (1990)
11. Lincoln, Y.S.: The making of a constructivist. In: Guba, E.G. (ed.) The Paradigm Dialogue, pp. 67–87. Sage, Newbury Park, CA (1990)

Chapter 4
Competitiveness Strategies for Negative Organizations

Abstract Chapter 4 is a long and comprehensive chapter. In it we discuss a number of important and central themes. The results of the interviews and focus group are discussed, as well as the results of the study in general. The cases of Portugal and of Google are focused on, as are competitiveness strategies. How may a firm become more positive? What does change management and the implementation of strategies involve? A series of steps may be followed, to ensure success as regards change and becoming a more positive organisation.

We currently live in a hyper-competitive environment due to the emergence of a global economy and rapid technological change [1]. Competing in environments with these characteristics requires organisations and their managers to apply the mindset needed to keep up with and anticipate changes, and it is therefore essential that they make decisions quickly and accurately [2].

The strategy that an organisation implements is fundamental to determining its success and competitiveness in the future. Strategy is understood as requiring choices [3] and commitments which need to be integrated and coordinated—involving actions designed to take advantage of core competences and hence achieve competitive advantage [1]. A strategy refers to the plan, implemented by the upper hierarchical levels of the organisation, in order to achieve the organisation's purpose [2]. For an organisation to be strategic, it must be prepared to make decisions, managing to build a unique position of value in the marketplace which involves a unique set of activities [4]. In addition, the implementation of strategies will foster efficiency and a more reliable anticipation of changes in the market and in other organisations [4].

According to Porter [4], there is a big difference between strategy and operational efficiency, because even though operational efficiency is relevant, it is not enough. For example, Japanese companies are recognised for being efficient, but most of them do not develop strategies that distinguish them from one another [4]. In addition, the competitive advantage of a product or service is limited, i.e., the time in which the organisation has a competitive advantage over its competitors and for that particular product or service, is determined by the time it takes its competitors to acquire the necessary skills to imitate it [1].

B. Graça and M. Au-Yong-Oliveira, *Competitiveness Strategies for Negative Organizations*, Information Systems Engineering and Management 1,
https://doi.org/10.1007/978-3-031-59914-9_4

In this sense, it is crucial for organisations to be able to respond to the various types of demand, as well as the opportunities that exist in highly competitive and uncertain environments. Organisations therefore need to be strategically flexible on an ongoing basis in order to have the capacity to manage the uncertainty and risks inherent in their various businesses [1].

The purpose of this chapter is to provide strategies that companies, governments and employees can adopt to make Portuguese organisations (and mainly other relationship-culture firms) more positive and competitive. To this end, the results obtained from the interviews and Focus Group will be presented below. Subsequently, the results will be discussed in Sect. 4.2 and then, in Sect. 4.3, strategies will be outlined that organisations, employees and governments can adopt to make them more positive. Finally, Sect. 4.3 explains how organisations should manage the process of change towards more positive organisations.

4.1 Results of the Interviews and Focus Group

As mentioned above, six interviews were carried out for this research in order to understand the perspective of the selected sample. Table 4.1 shows the questions asked of each of the interviewees. Each of the interviews will be explained in more detail in the following subchapters.

In addition to the interviews conducted and analysed in detail in Sects. 4.1.1, 4.1.2, 4.1.3, 4.1.4, 4.1.5 and 4.1.6, a Focus Group was also held, which will be explained in more detail in Sect. 4.1.7.

4.1.1 Thematic Analysis of the Interview with João Ranito

The initial conversation between interviewee João Ranito and the authors of this research took place on 18th October 2022. In addition to this initial conversation, João Ranito responded in writing to 10 questions proposed by the first author, which were later received by her on 20th October 2022. A brief contextualisation of the interviewee—João Ranito—can be found in Table 4.2.

The initial conversation, which lasted around an hour and sixteen minutes, was aimed at getting to know the perspective of an innovator and entrepreneur on the impact of a negative organisation on its competitiveness.

Through the questions mentioned in Table 4.1, the intention was to understand the interviewee's point of view, in the position of manager and entrepreneur, with regard to the questions posed. Thus, as the literature review revealed, the concepts of negative organisation and positive organisation proved to involve various dimensions for the interviewees, particularly their ability to grow, the financial sustainability of the business and their ability to "see the future", i.e., their ability to understand how

Table 4.1 Identification of the questions asked in the interviews conducted

Question	Theoretical foundation	Interviewee
How do you measure the competitiveness of a project?	Measuring an organisation's competitiveness is increasingly important. As organisations today operate in dynamic, global markets, it is important to understand, from the interviewee's perspective, how organisations can tell if they are competitive The competitiveness of a project and an organisation is subjective and involves several dimensions, so it doesn't just involve a single factor. To this end, we sought to understand the interviewee's perspective on the metrics, indicators and approaches used to assess and measure the sustained competitiveness of projects and/or organisations	João Ranito
What influence does innovation have on an organisation's competitiveness?	As evidenced by the literature review, the importance of innovation for an organisation's competitive future is reiterated [5]. It is therefore important for organisations to be able to adapt to changes in the ecosystem in which they operate [6] In this sense, the question aims to explore the interviewees' perspective on innovation, both incremental and disruptive, and how this influences the organisation's competitiveness	João Ranito Issam Ghazzawi Veronica Perez Frank Fialho
What influence does organisational culture have on competitiveness?	Organisational culture plays a significant role in the ability of organisations to be "ahead of the curve" and thus be able to anticipate possible paradigm shifts and, consequently, plays a relevant role in ensuring an organisation's competitiveness. [7] It is therefore important to understand how the values, beliefs and practices shared by all the organisation's employees shape the way they behave and their performance, as well as the strategies implemented. All these aspects naturally influence the organisation's competitiveness and its competitive advantage in a dynamic, global environment [1] In addition, the interview with the former Google employee was intended to understand how the organisational culture has led Google to become a disruptive, innovative and successful organisation on a global scale. It is also important to note that having a competitive advantage at a given moment does not guarantee its perpetuation—it is essential to foster an organisational culture that continually promotes discomfort, creativity and innovation, thus avoiding stagnation [8]	João Ranito Tommy York Issam Ghazzawi Veronica Perez Frank Fialho

(continued)

Table 4.1 (continued)

Question	Theoretical foundation	Interviewee
What strategies do you adopt to ensure your organisation's competitiveness?	Ensuring the competitiveness of an organisation is a very complex and important role for all leadership teams [9] In this sense, we are trying to understand what strategies the interviewee uses to ensure that the organisation is competitive	João Ranito
What do you see as the government's role in ensuring that Portuguese organisations are more positive and competitive?	The government has a central role to play in promoting more positive and competitive organisations [10]. The government should act as a facilitator by creating conditions that encourage investment in research and development, as well as contributing to collaboration between universities and private companies [11] In this sense, understanding how government policies can help organisations to be more competitive is essential, especially as organisations operate in a global market	João Ranito
What do you think a negative organisation is? What do you think a positive organisation is?	Given the complexity of defining the concepts of "positive organisation" and "negative organisation", it was considered important to ask the interviewee's opinion as a Portuguese entrepreneur In addition, the definitions of positive and negative organisations, as well as being subjective, as was seen in the literature review, are concepts that have been little explored, but which are at the core of this research It is therefore important to understand, from the interviewee's point of view, which characteristics make an organisation positive or negative	João Ranito
What strategies should employees adopt to make the organisation more positive?	Employees are the cornerstone of every organisation and are therefore essential to ensuring that the organisation they are part of is a more positive one [2] In this sense, it is important to understand how they can actively contribute to building a more positive organisation	João Ranito

(continued)

Table 4.1 (continued)

Question	Theoretical foundation	Interviewee
What are the strengths and weaknesses of working for a company like Google?	The interview began with the aim of understanding the interviewee's overall view of what it was like for him to work at a technology giant like Google. In this sense, and as discussed in Sect. 2.1.1 of this study, it was possible to see that an organisation is made up of different dimensions, so there are positive aspects in an organisation that is mostly negative, and vice versa With regard to Google, a company perceived as a positive organisation by both external and internal audiences (Comparably, 2023) the present question aimed to ascertain whether the reality exposed by the Literature Review was correct, i.e., whether this organisation is also made up of positive and negative dimensions	Tommy York
After the explanation given between a negative and positive organisation, how would you frame Google and the team you were part of? Do you think there is any discrepancy? Do you think the same thing that happened to Kodak could also happen to Google?	Once again, it is clear that an organisation, whether positive or negative, cannot be defined by a single dimension. The interviewee emphasised the transparency in the selection and promotion of employees at Google, but pointed out flaws in this process that make it less fair. Promotion at the company is based on the launch and adoption of new software, neglecting to recognise those who maintain existing applications. This system fosters intense internal competition, jeopardising collaboration between different teams. Google needs to align its career progression model with global objectives and promote collaboration in order to differentiate itself from competitors [5] History reveals that there are several examples of organisations—for example Nokia, Kodak and IBM—that have lost their competitive advantage and market share due to their internal policies [6]. In this sense, the launch of a new and innovative work tool—GPT Chat—by a Google competitor shows that Google's current position in the market may be in decline and heading towards failure	Tommy York

(continued)

Table 4.1 (continued)

Question	Theoretical foundation	Interviewee
The differences between the cultures of different countries naturally influence the culture of organisations, as well as the way they work. For example, compared to the US, China is a country that sees relationships as the best metric for promotion qualification. Do you think that a country's culture can influence organisations to be more positive or negative?	Each country and culture have their own own values and tendencies, which naturally influence individual behaviour, including in the work environment and the way they work [12] The aim of this question was to find out whether the Culture factor has an impact on the competitiveness of organisations	Tommy York Issam Ghazzawi Veronica Perez Frank Fialho
Based on your professional experience, how do you think we should measure the negativity of a country and an organisation?	Considering that an organisation is made up of multiple dimensions that can influence its negativity or positivity, it is crucial to understand how we can measure this negativity in order to be more attentive and develop strategies that will turn negative organisations into positive ones in the future. It is in this context that this question arises	Tommy York Issam Ghazzawi Veronica Perez Frank Fialho
Do you think we should resign ourselves to the *status quo* or should we fight against the negative points of an organisation?	Change is present in every individual and, although every employee has the opportunity to alert the organisation to less positive situations, it is not uncommon to face obstacles when the organisation's interests are aligned with maintaining the current paradigm. As a rule, individuals have a lot to lose and the organisational culture may not always receive criticism or suggestions for improvement in a way that is positive [13] It is therefore important to understand, from the perspective of the interviewees, whether employees should fight against the *status quo* and whether they consider it important to have a flexible and open organisational culture so that employees feel safe to express their opinions in a respectful and justified manner	Tommy York Issam Ghazzawi Veronica Perez Frank Fialho

(continued)

Table 4.1 (continued)

Question	Theoretical foundation	Interviewee
Do you think that having a negative person ("bad apple") in the team can demoralise and demotivate the rest of the team and, consequently, have an impact on the team's competitiveness?	The analysis of the literature revealed that the team and its work have a major impact on an organisation's competitiveness, and thus the behaviours and actions of the various team members determine its success In this sense, having just one destabilising element can have emotional consequences for the rest of the team and, in turn, impact on the organisation's productivity [14] The question was therefore intended to find out whether, according to the professor's perspective, he thought that people who adopt negative behaviours would have an impact on the organisation's competitiveness	Issam Ghazzawi
Corporate psychopaths adopt charismatic, charming and manipulative behaviour to disguise their true nature. How do you think we should identify these individuals and how can we "combat" them?	The literature has revealed the presence of psychopaths in organisations—Corporate Psychopaths. It is therefore important to see how employees and managers can be aware of these behaviours and identify these individuals. Consequently, the organisation will know how to deal with these problematic individuals with everyone's well-being in mind and then apply the appropriate measures [14, 15]	Issam Ghazzawi
Apart from organisational culture, what other factors influence an organisation's competitiveness?	Throughout the study of the subject under analysis, it was possible to see that Organisational Culture has a major impact on an organisation's competitiveness [12]. However, and especially because organisations do not operate in isolated ecosystems, it is also important to identify other key factors for their competitiveness and success [7, 16]	Issam Ghazzawi Veronica Perez Frank Fialho
What strategies should employees adopt to make the organisation more positive? Should the strategies come from employers or employees?	Although organisational culture is defined by the measures and behaviours adopted by the organisation's top leadership, employees also play a key role in its dissemination. It is therefore important to understand how employees can contribute to creating a more positive organisation	Issam Ghazzawi Veronica Perez Frank Fialho

(continued)

Table 4.1 (continued)

Question	Theoretical foundation	Interviewee
How can strategic marketing help an organisation to be more competitive?	Given the abundance of information in all media, be they traditional media or the Internet, the way in which the product is publicised and the platforms used to publicise it play a key role in the exposure of the product to the consumer, and consequently the way in which the product is sold, the turnover and the prominence of other companies The aim of this question was to understand the interviewees' perspective on the proliferation of strategic marketing, whether digital or analogue, as a tool for selling and promoting products, and therefore its influence on competitiveness	Veronica Perez Frank Fialho
Do you think it is important to be resilient in Portugal?	Resilience is crucial due to the need to overcome persistent obstacles, adapt to economic and social changes, and face structural limitations. Thus, through this question, it is possible to ascertain whether resilience plays an important role for the interviewee	Anonymous participant
Do you think it is easier to innovate in large companies or in small companies/ start-ups?	The question aims to understand the dichotomy between innovation in large companies and in small companies or start-ups, as well as the advantages and disadvantages in each case	Anonymous participant
Do you think there is a problem in Portugal with low salaries?	This question aims to clarify whether the interviewee considers that Portugal's low wages, when compared to other European Union countries, are a factor that contributes to the country's lack of competitiveness	Anonymous participant
What is the most important factor for a leader's success—motivation, talent or infrastructure?	Motivation is common to all successful leaders, but we wanted to understand whether talent and infrastructure also play a central role	Anonymous participant
Do you think motivation is intrinsic?	The question is whether motivation is born with each individual or whether it is something that can be nurtured and fostered	Anonymous participant
How envious are the innovators and successful people in Portugal?	As Portugal has socio-economic challenges and an organisational culture of relationships, and envy can sometimes become a problem for leaders	Anonymous participant
Do you know what negative organisations are?	This question shows the interviewee's perspective on the concept of negative organisations	Anonymous participant
Do you think there are more negative organisations in countries with relational cultures than in countries with transactional cultures?	Given the cultural differences between countries, this question aims to understand whether these differences have an influence on organisational culture and, consequently, on the number of negative organisations	Anonymous participant

Table 4.2 Identification of interviewee—João Ranito

Name	Provenance	Academic training	Function
João Ranito	Porto, Portugal	Faculty of Engineering, University of Porto M.Sc., Genetic Algorithms 1991–1993 Faculty of Engineering, University of Porto B.Sc 1983–1988	Manager, entrepreneur and former senior manager (Novabase)

their business and organisation will evolve and how they can implement changes that are in line with that future.

It was therefore found that, from their perspective, organisations should have a strong focus on the future and how they can be more competitive. In this sense, only by focusing on the future and future competitiveness can organisations be successful. Change and the way organisations manage this period is essential in order to keep up with the growth of the business itself, as well as the organisation itself. It is therefore essential for organisations to be agile enough to keep up with market changes and to be sustainable at all levels.

In this sense, the initial conversation resulted in the following main ideas, in Table 4.3.

Strategies for competitiveness: As can be seen from the literature review developed and explained above, the strategies adopted by organisations have a profound impact on current and future competitiveness. According to the perspective of the interviewee—João Ranito—it was possible to highlight some of the strategies that organisations could implement in order to be more successful. Some of these strategies include:

(i) hire better professionals and provide them with a career that is attractive to their abilities and personal ambitions;

Table 4.3 Main themes discussed during the interview with João Ranito

Topic under discussion	Topics covered
Strategies for competitiveness	– Critical factors for the success of organisations – The Role of Leadership, Government and Employees
Importance of change	– Adaptability – Agility – Investment in Innovation
Organisational culture and competitiveness	– Impact of Organisational Culture on Competitiveness
Growth and sustainability	– Balance of results – Effective strategies

(ii) attention by the organisational structure to the real needs of current and potential customers;
(iii) clearly define the organisation's core business and focus your efforts on core-related goals;
(iv) developing a network of partners that complements the organisation and is aligned with its objectives;
(v) promote a culture of quality and pride in the work that is done;
(vi) create a decentralised structure in which each leader has decision-making autonomy, within the limits imposed by the overall objective and the values/ culture of the organisation.

Employees play a crucial role in the operations of all organisations, contributing significantly to their success and dynamics. They carry out tasks, make decisions and interact with customers, suppliers and co-workers. As the driving force behind organisational operations, the way in which employees engage with and contribute to the working environment is extremely important. The interviewee highlighted the relevance of specific measures that can be adopted to promote a more positive and effective organisational culture:

- Firstly, the interviewee suggested that employees don't hesitate to voice their concerns and disagreements when they realise that something isn't working properly. This attitude not only demonstrates a commitment to continuous improvement, but also allows issues to be identified and addressed proactively, preventing problems from escalating or perpetuating.
- In addition, the interviewee emphasises the importance of cultivating a sense of pride in their work. The appreciation and satisfaction of employees in their work activities becomes a direct reflection of their productivity, attitude and interactions with colleagues. Fostering an environment in which employees recognise the positive impact of their contributions can increase motivation and cohesion within the team.
- Finally, the interviewee emphasises the need for employees to seek a career that meets their personal aspirations. By aspiring to a more rewarding career path in line with their individual goals, employees not only maximise their own potential, but also contribute to the overall dynamics of the organisation. This approach can positively influence the organisational culture, encouraging evolution and the pursuit of excellence.

To summarise, the suggestions put forward by the interviewee aim to encourage employees to play an active role in creating a more positive and productive working environment. By encouraging the expression of opinions, promoting pride in one's work and encouraging the pursuit of a meaningful career, organisations can boost the commitment and positive impact of their employees.

In addition to the influence of the strategies adopted by the organisations themselves, the government and government policies play a key role in developing strategies that boost and contribute to their competitiveness and, consequently, that of the country itself. Thus, understanding the role of government policies in promoting the competitiveness of organisations is of great importance, particularly as organisations operate in a global market. According to the interviewee, the Portuguese government should adopt a set of strategic measures aimed at strengthening the competitiveness of national companies:

- Firstly, it is suggested that the government work on building a positive image of the country by promoting well-founded domestic initiatives. This will help to open up the international market, encouraging organisations to become more competitive by associating themselves with a positive image of the country.
- In addition, the government should identify strategic areas for investment and explore opportunities for co-investment in partnership with private companies. This approach can boost growth and innovation in sectors considered key to economic development.
- The comprehensive application of technologies in the public sector is a priority, as it can optimise the efficiency and agility of government processes. Ensuring that the public sector acts as a driving force, rather than an obstacle, is essential to fostering overall competitiveness.
- Promoting the public purchase of national solutions, in accordance with free competition regulations, is a way of supporting local companies and stimulating innovation at the national level.
- Making legislation more flexible, especially with regard to the labour market, in order to adapt to contemporary realities, is highlighted as a crucial measure to enable organisations to be agile and responsive to change.
- Continuous investment in the training of individuals of all age groups is considered essential. The competitiveness of organisations is intrinsically linked to agility, which in turn is driven by comprehensive and up-to-date knowledge.
- Finally, streamlining the justice system to ensure efficiency is emphasised. Effective justice contributes to competitiveness by avoiding situations that could give negative organisations a competitive advantage.

In summary, the interviewee's suggestions indicate that government policies play a central role in creating an environment conducive to the competitiveness of organisations. By promoting a positive image of the country, investing in strategic sectors, adopting innovative technologies, supporting continuous training, adapting legislation and optimising the public sector, the government can contribute significantly to the development of a favourable scenario for the growth and success of organisations.

Importance of change: The ability to adapt regularly is a fundamental principle for ensuring the competitiveness of organisations, especially in a constantly evolving business environment. To accommodate the growth of the organisational structure and meet the challenges that arise, agility becomes a crucial necessity. Constant change is a reality and, in order to remain competitive, organisations need the flexibility to adjust to new circumstances and opportunities.

This need to adapt is directly related to the importance of innovation for the future competitiveness of organisations. The ability to adapt to changes in the business ecosystem is essential to maintaining an advantageous position in the market.

As the interviewee emphasised, the terms "change" and "innovation" are intrinsically linked. The context of the organisation's life cycle and the market influences the approach to innovation. Organisations can choose an approach of (i) continuous innovation, adjusting incrementally to optimise their position, or (ii) disruptive innovation, characterised by significant technological advances. Especially in sectors such as Information Technology (IT), innovation plays a critical role in success.

Furthermore, the ability to adapt and the importance of innovation are strongly influenced by organisational culture. This culture plays a vital role in determining the right balance between "agility" and "optimisation", not only defining the way the organisation operates, but also shaping employees' attitudes and values towards change and innovation.

Therefore, it is the effective combination of all the above factors—change, innovation, agility and optimization—that ensures an organisation's competitiveness. The ability to embrace regular change, recognising the importance of innovation and cultivating an organisational culture that supports agility and adaptation is essential for thriving in a dynamic business environment. Only when these elements work together will an organisation be prepared to challenge the *status quo* and position itself advantageously for the future.

Organisational culture and competitiveness: A fundamental step in guaranteeing an organisation's competitiveness is having a leadership and management team capable of playing the necessary roles. This includes defining a clear focus for the organisation and ensuring that all employees fully understand their roles within the organisation and the extent to which they are key to its overall success. The role of ensuring the competitiveness of an organisation is of paramount importance and complexity and is central to all leadership teams. Although this task may vary from one organisation to another, some common strategies can be adopted. One of these is the creation of a decentralised structure, in which each leader has the autonomy to make decisions within the limits set by the global objectives and the values and culture of the organisation.

Growth and sustainability: According to João Ranito, a positive organisation is one that manages to promote its own growth and sustainability. This means balancing its results fairly between all stakeholders, such as shareholders, employees, customers and other partners, while reserving the necessary resources for its continued development. A positive organisation has a clear vision of its future and implements measures in the present to ensure its future competitiveness.

In contrast, a negative organisation takes the opposite approach. It focuses on the present, jeopardising its future and showing a reluctance to improve. What is more, it does not distribute its results evenly among its various stakeholders, jeopardising long-term sustainable development.

Thus, the presence of a competent leadership team, the adoption of effective strategies and the promotion of a positive organisational culture are crucial factors in ensuring an organisation's competitiveness. These actions not only contribute to its present growth, but also lay the foundations for a competitive and sustainable future.

We can therefore conclude that the interview with João Ranito resulted in the identification of the following themes and patterns—please see Table 4.4.

4.1.2 Thematic Analysis of the Interview with Tommy York

On 30th March 2023, an interview was conducted with a former Google employee by the research team/book authors. Table 3.3 shows a brief contextualisation of the interviewee. The main aim of this interview was to get the employee's perspective on whether Google—a technology giant—would fall within the spectrum of a positive or negative organization (Table 4.5).

The interview lasted approximately two hours and the questions listed in Table 4.1 were addressed.

Tommy York was one of the employees included in Google's collective redundancy process in January 2023 [17]. The context of his dismissal comes at a negative time in Tommy's personal life, as he had only been back at work for three days. His mother was diagnosed with (stage 4) pancreatic cancer in November 2022, and he felt the need to support his mother at this difficult time. In agreement with his employer, he took family support leave when his mother's health deteriorated and, weeks later, when his mother was unable to resist the illness, Google granted him four weeks' leave (Bereavement Licence) to mourn and rest.

Tommy demonstrated that there was a feeling that they chose the people who would be included in the collective redundancy at random, without looking at the reasons that led to their absence, i.e., according to his perspective, Google did not only take into account professional performance factors. The collective redundancy list included employees who, like Tommy, needed to take some kind of leave permitted by Google—for example (i) leave to support family members, (ii) leave for the bereavement of a loved one, (iii) pregnant women who would need to take leave soon or women who were already taking such leave and (iv) other types of leave, such as sick leave. In addition to these elements, employees were also included who are immigrants with a visa and who needed to find a job in their professional field within a maximum of 60 days, and if they did not, they would be forced to return to their country of origin.

The main topics covered in the conversation with Tommy York are presented below. Table 4.6 highlights the most important topics, followed by the theoretical justification for them.

Focus on Corporate Wellbeing: Google, a technology giant based in Silicon Valley in the United States, prides itself on being a company that cares about its employees and their wellbeing and has even been highlighted as one of the "100 Best Companies to Work For". [8]. In this sense, Google offers a wide range of benefits to its employees. As you can see on its official website (Google [18]) these are focused on five main categories:

Table 4.4 Themes and patterns identified—João Ranito interview

Themes and patterns identified	Quotes to justify this evidence
Strategies to ensure the competitiveness of organisations	During the interview with João Ranito, it was possible to see how the entire ecosystem in which organisations operate influences their competitiveness. In this sense, the interviewee suggests complementary strategies that should be adopted by Organisations, Government and Employees Strategies for Organisations – Human Resources Management: hiring the best candidates, giving them continuous training, an ambitious goal and allowing them to make their way, rewarding them in a way that is aligned with the organisation's strategic objectives – Having an organisational structure that is attentive to the current and future needs of consumers and, in this sense, puts pressure on the organisation to act accordingly – Define the company's core business and focus efforts on this objective, outsourcing to partners everything that is not part of the organisation's core business – Create a network of partners that complements the company's core competencies and ensures that these stakeholders are aligned with the organisation – Ensure that all people grow and learn and promote a culture of quality and pride in the results of their work – Create a decentralised structure in which each leader has decision-making autonomy, within the limits imposed by the overall objective and the values/culture Government Strategies – Create a positive image of the country by investing and publicising good internal initiatives, thus opening up the international market and encouraging organisations to be more competitive – Define strategic investment areas and co-invest with companies – Apply all technologies in the state apparatus to streamline the public sector. It is also important to ensure that the public sector is part of the country's engine, not its brake – Promote the public purchase of national solutions, as other countries do, within the legal constraints of free competition – Make legislation more flexible, adapting it to the new reality, including everything that concerns the labour market – Invest in training people of all ages. Competitiveness depends on agility and agility depends on the breadth of knowledge – Speed up justice by ensuring that it is effective. When it is not, negative organisations can even have a competitive advantage Strategies for Employees – Protest when they feel that something is not working properly – Invest in a sense of pride in their work – Fight for a better career that fulfils them individually

(continued)

Table 4.4 (continued)

Themes and patterns identified	Quotes to justify this evidence
Influence of organisational culture and leadership style	Organisational culture plays a key role in ensuring an organisation's competitiveness. Thus, according to the interviewee's perspective, leadership and the leadership style adopted allow the values and behaviours that govern the organisation and its employees to be disseminated The interviewee also pointed out that the organisational culture also determines the comfortable balance between agility and optimisation
The importance of innovation	Organisations must be able to adapt to changes in the ecosystem in which they operate. According to the interviewee, change and innovation are correlated terms. The life cycle of the organisation and the market naturally influence the type of focus on innovation, i.e., the organisation can focus on (i) continuous innovation, making small adjustments to optimise its position in the market or (ii) disruptive innovation, usually characterised by a technological breakthrough. It is important to emphasise that for IT companies, innovation is a critical success factor
Differences between negative and positive organisations	For João Ranito, a positive organisation is capable of promoting its own growth and sustainability, managing to balance its results between all stakeholders (i.e., shareholders, employees, customers and other partners), while safeguarding the resources necessary for its growth. In this sense, a positive organisation is able to see clearly what its future will look like and is therefore able to implement measures in the present that will enable it to be competitive in the future. On the other hand, a negative organisation is the opposite, since it "lives in the present, mortgages the future, doesn't want to improve, and doesn't distribute its results evenly among all *stakeholders*"
Strategies for measuring a project's competitiveness	According to the interviewee, and especially for companies in the Information Technology (IT) sector, it is important for the company to be competitive: – Have a robust market share (at least the segment average), as well as other macro factors (such as turnover, profit margin and growth potential); – Sell more products than they produce; – Have enough money to support the investment of new hires; – Change regularly in order to accommodate the growth of the organisational structure, i.e., be agile; – Have a leadership/management team with the capacity to fulfil the necessary roles, including setting the focus for the organisation and ensuring that all employees understand their role in it and how they are integral to its success; – Ensure that the organisation's offer is aligned with market needs; – Have a clear understanding of who the organisation's key partners are and manage them efficiently

Table 4.5 Interviewee identification—Tommy York interview

Name	Age	Provenance	Academic training	Function
Tommy York	33 years old	San Francisco, California	University of California, Berkeley BA, East Asian Languages and Cultures—Chinese 2007–2012 University of California, Berkeley, Haas School of Business BSc, Business Administration 2010–2012	Software Engineer

Table 4.6 Main themes discussed during the interview with Tommy York

Topic under discussion	Topics covered
Focus on corporate wellbeing	– Work-Life Balance at Google – Level of Turnover and Absenteeism – Benefits offered by Google (including salary)
Innovation	– Stagnation and possible threats to Google's competitive future (e.g., Chat GPT) – Creativity Bias and the internal employee reward and evaluation system
Organisational culture and competitiveness	– Career progression; – The impact of organisational culture on competitiveness

Health and wellbeing

- Medical, dental and eye insurance for employees and dependents;
- Employee assistance programmes focused on mental health;
- Workplace accommodation for physical or mental health problems;
- Wellness centres in the workplace;
- Access to mental health applications;
- Second medical opinion for you and your loved ones;
- Medical defence programme for transgender employees.

Financial well-being

- Competitive remuneration;
- Regular bonuses and opportunities to upgrade shares;
- Generous 401(k) retirement plans and regional plans;
- Analysis and annual adjustments of pay equity between companies;
- Repayment of student loans;
- Individual financial coaching.

Flexibility and free time

- Paid time off, including holidays, bereavement, jury service, sick leave, parental leave, disability and public holidays;
- Hybrid working model with remote working opportunities also available;
- Four weeks of "work from anywhere" per year;
- Part-time work and job-sharing options.

Family support and assistance

- Support for fertility and family growth;
- Parental leave and baby bonding leave;
- Leave for carers;
- Support and care for the elderly;
- Childcare assistance;
- Survivor's income allowance.

Community and personal development

- Educational reimbursement;
- Googler-to-Googler training platform for colleagues;
- Matching donations and free time for volunteering;
- Employee resource groups;
- Internal Googler community groups and local culture clubs.

From the interviewee's perspective, his career at Google was marked by many positive aspects, and the benefits offered were a major part of this result. However, the same organisation was ruthless in firing Tommy just three days after he returned from leave due to his mother's death. During the difficult period he went through, his employer was excellent, since, as already mentioned, Tommy had the opportunity to look after his mother in her last days, to mourn her, as well as having internal support groups at his disposal that gave him the strength he needed to get through it. However, Tommy wonders what the point of these benefits is if in the end they sack employees without giving them a reason.

In this sense, management practices such as collective redundancies and restructuring have a major impact on the physical and mental well-being of both the employees who are made redundant and those who remain employed. Although there is no evidence that collective redundancies or downsizing have a positive impact on the organisation's finances, many companies resort to this practice. In fact, collective redundancies are associated with a negative stock market return and downsizing is associated with a negative effect on market returns, as well as increased fear and lack of trust in the workplace [20]. It is important to realise that a lack of trust in the organisation's management can have significant consequences for competitiveness, such as (i) employees may miss more work, (ii) the quality of their work may be compromised and (iii) important deadlines may be missed [19].

The dismissal of an employee can generate a period of stress for the individual, and in the United States, for example, as in the case of Tommy York, it often results in the loss of health insurance, jeopardising the physical and mental well-being of those affected, since the national health system represents a high cost that cannot be borne by the user [20].

Innovation: Google is recognised as being an innovative company, having over the last few decades solidified its position in the competitive and volatile market in which it operates, as it has a strong focus on research and development of new products. In addition, Google has launched a series of innovative products and services - such as the Google Search Engine, Android (mobile operating system), Google Maps, Google Assistant (virtual assistant), Google Photos, Google Cloud Platform and Google Pixel (line of smartphones), which have fuelled its growth and further solidified its position in the global market.

However, despite its undisputed position as one of the main market leaders, according to the interviewee, this position may be vulnerable. As outlined in Sect. 4.2.1, the current context is extremely volatile, requiring organisations to constantly reinvent themselves in order to remain relevant to their target audience. Given this need, the interviewee expressed concern about Google's future, given that the company's technology has not undergone significant reformulation in the last 10 years, particularly the Google Search Engine. From his perspective, Google remains the market leader solely due to the lack of a competitor capable of rivalling it. However, the current lack of competition does not necessarily mean that, in the future, a competitor cannot be more effective and overcome the competitive advantage currently held by Google.

There are countless examples of organisations that have not been able to adapt to new market paradigms and have therefore been unable to maintain their competitive edge. Companies like KODAK, IBM and Nokia are characterised by having developed an "immune system" that fights off any products or services that don't correspond to their core business. In doing so, they develop a "marketing myopia" that prevents them from foreseeing structural changes in their business and make poor decisions that jeopardise the sustainability of the organisation [7] as they insist on continuing to exploit the product or service that has given them a long-standing advantage [6]. These companies have failed because innovative competitors have emerged in the same market, who, even though they have similar products, have innovated. The same could happen to Google if a "rival" emerges that innovates in the services provided within the same market that Google operates in. It is therefore crucial that organisations accept the idea that their once-successful business models need to be changed in order to maintain their competitive advantage [6].

However, as the technology industry is constantly changing, it is difficult for organisations in this sector to anticipate all the paradigm shifts and Google is no exception. One of the main concerns for the future of both Google and other organisations in various industries and sectors is represented by GPT Chat, an application funded by Microsoft, which has the potential to revolutionise the way searches are carried out and information accessed, which could significantly impact one of Google's main sources of revenue. In response to this threat, Google is already in the process of developing a potential alternative to this innovative technology, called Google Bard. Currently, this tool is only available in 180 countries and is supported by three different languages: English, Japanese and Korean [21]. With this tool in particular, Google, in order to remain comfortable with the position it occupies, has stopped innovating and has moved into the passenger seat.

In addition to the threats mentioned above, Tommy York highlighted some internal problems that, in his opinion, could affect the innovation process at Google. These problems include:

- The need for Google employees to familiarise themselves with and program using the company's proprietary software. As a result, the first six months of working for the organisation are mainly dedicated to learning the various softwares used. Although this software is not complex, the time needed to learn this new programming language requires new employees to dedicate the first few months to learning it.
- Tommy York also highlighted the presence of a creativity bias within the organisation, since those who create products or services are rewarded more than those who work to maintain existing products or services, even though this work is fundamental to the organisation's success. Launching a product and having visibility is therefore valued within the organisation, while maintaining an existing service, such as Gmail, is not considered as attractive to the team of software engineers. However, employees feel that Google often cancels new and innovative projects, even after several hours of work dedicated to developing these projects. The interviewee mentioned that it is not beneficial to work for a company that constantly cancels projects, as this can damage the individual's career trajectory.

Organisational Culture and Competitiveness: Organisational culture has a significant influence on the competitiveness of companies, since the ability to adapt and anticipate change determines their success or failure. As mentioned above, Google values the well-being of its employees and endeavours to implement internal policies that enhance their work-life balance. In Tommy York's view, organisational culture and internal policies are the main reason why Google has not introduced disruptive products or services.

Currently, there are several Google applications that have not undergone any structural or innovative changes. One example is Google Maps, where the visualisation differs depending on the device used. If you access it from your computer, you can see the map, but if you access it from a mobile device (such as a smartphone or tablet), you can no longer see it. Another example is Google Inbox, which was an application planned to replace Gmail. However, the project was cancelled without an explanation being given to consumers. Some of the employees involved in the project were relocated to Gmail and some of the features of Google Inbox were incorporated into this application.

Tommy York also points out that companies like Amazon, Facebook, Apple and Google offer attractive salaries to their employees, which keeps these professionals working for these large companies. Many of these employees are highly creative people who, if they were in start-ups, would probably develop products or services that would compete with the products of these companies that have been innovating less and less.

From Tommy's perspective, Google maintains its position as market leader because there is still no competitor to match. He argues that Google and other big tech companies are currently looking for stability, so they do not prioritise innovation.

It can therefore be concluded that the interview with Tommy York resulted in the identification of the following themes and patterns—please see Table 4.7.

Table 4.7 Themes and Patterns identified—Tommy York interview

Themes and patterns identified	Quotes to justify this evidence
Benefits and concern for employee well-being	Tommy highlights the benefits offered by Google, such as comprehensive medical insurance, mental health assistance programmes, fertility support and family care, and educational reimbursement. These benefits are considered important for the well-being of employees. In addition, Google pays more than other companies in the Bay Area (San Francisco), so the salary Google pays makes it sustainable to live in a city like San Francisco
Support during difficult times	Tommy highlights the support he received from Google when he needed to look after his mum during her illness and bereavement. The company offered leave and internal support groups, showing concern for the well-being of employees in challenging personal times
Sudden dismissal and lack of transparency	Tommy reports that, even after receiving support in difficult times, he was dismissed without a clear or reasoned justification. This experience raises questions about the company's behaviour when it comes to dismissing employees. Tommy York was also concerned about the emotional and financial impact this dismissal had on him
Innovation management, competitiveness and digital transformation	Google's stagnation and the threat posed by Chat GPT highlight the importance of managing innovation and competitiveness in organisations. As a company known for its innovative culture, Google faces the challenge of maintaining its competitive edge and avoiding stagnation. This requires effective strategic management, including identifying new opportunities for innovation, stimulating creativity and developing agile approaches to adapt to market changes
Attractiveness of Human Resources	Tommy points out that Google is progressively attracting human resources with more professional experience by offering attractive benefits. On the other hand, Google's organisational culture no longer seems to have innovation at its core, so employees have less creative freedom. More experienced employees, as a rule, tend to look for jobs with greater stability and that offer a better benefits package

4.1.3 Thematic Analysis of Interview with Prof. Dr. Issam Ghazzawi

The interview with Professor Issam Ghazzawi took place on 14th September 2023. The main aim of the conversation was to understand the Professor's perspective as a professional with extensive experience in the field of Strategy and Leadership. A brief contextualisation of the interviewee can be found in Table 4.8.

The conversation with the teacher lasted around an hour and the questions listed in Table 4.1 were addressed. As with the interviews above, you can also see in the same table a theoretical justification for the choice of questions.

During the conversation with Professor Issam Ghazzawi, it became clear that, from his perspective, the leadership of an organisation has a major impact on its competitiveness. Therefore, without a strong organisational culture, employees will not be as committed to it.

Table 4.9 presents the main topics covered, followed by a theoretical justification for them.

Organisational Culture and Competitiveness: According to the Professor, there are several factors that influence whether an organisation is more positive or negative, namely its leadership and the way it operates, as well as the culture of the country in question.

If, on the one hand, an organisation with a positive culture sees its employees as a single team, organisations with a negative organisational culture usually have a greater challenge in terms of their future survival. According to the interviewee, in a negative organisational culture, employees are not committed to the organisation and periodically consider whether they should leave, which, in the long term, naturally damages the organisation's competitiveness. Therefore, for the Professor, an organisational culture should be able to stimulate creativity, innovation and teamwork.

In addition, and with the organisation's competitiveness in mind, it is important that it manages to be consistent in dimensions such as:

(i) research and development of new products;
(ii) establishing an innovative team;
(iii) working capital capacity (cash flow);
(iv) a strong and capable marketing team that efficiently manages the product's maturity cycle and understands how to introduce the product to the market;
(v) ability to rejuvenate.

In addition, the interviewee reiterates that in a positive organisational culture, there is no hidden conflict dimension, i.e., conflict is resolved and managed openly as it has a major impact on the organisation's productivity, as it confuses and distracts employees from their work. This dimension could also be one of the tools used by corporate psychopaths as a way of climbing and progressing within the organisation, one of the elements of a negative organisation.

Table 4.8 Identification of interviewee—Prof. Dr. Issam Ghazzawi

Name	Age	Provenance	Academic training	Function
Dr. Issam Ghazzawi	71 years old	Claremont, California	Ohio State University Master's degree, Labour and Human Resources 1980 Sul Ross State University Master's degree Business Administration (MBA) University of Pittsburgh Ph 1981–1982	University Professor, Management—University of La Verne

Table 4.9 Main themes discussed during the interview with Prof. Dr. Issam Ghazzawi

Topic under discussion	Topics covered
Organisational culture and competitiveness	– Commitment – Competitiveness – Corporate psychopaths
The key role of leadership	– Prevention of negative individuals—bad apples – Metrics to identify the type of organisation (negative or positive) – Strategies to make the organisation more positive
Conflict resolution and communication	– Address and resolve the conflict

As the literature review shows, Corporate Psychopaths are manipulative and use bullying as a weapon to achieve their personal goals. However, according to the teacher's perspective, the blame for the existence of these individuals in the organisation lies with the culture spread by the organisation's leadership. A good leader can easily see who these individuals are and take firm action against behaviour that could impact the organisation and other employees.

The primary role of leadership: Having a negative element in the organisation that is capable of demoralising and demotivating the rest of the employees depends essentially, according to the Professor's position, on the culture of the country and the organisation itself. For example, according to the Professor's perspective, these elements would not be tolerated in North American cultures, so verbal and written warnings would be issued and, if these did not work, the element would be fired.

So, according to his perspective, organisations have to be firm with these individuals—bad apples—and not allow these behaviours to be perpetuated in the future, because if they do not, the organisation's competitiveness will be at risk, since good and valuable employees could leave, especially if they see the leadership allowing these behaviours to be perpetuated.

Based on his professional experience, the professor believes that there are various metrics that organisations can use to determine whether the organisation is negative or positive. These metrics relate to (i) employee stress, (ii) employee turnover, (iii) employee commitment to the organisation, (iv) pride in belonging to the organisation, (v) employee absenteeism, (vi) employee performance and involvement in the organisation and (vii) inclusion.

In addition, there are strategies that the organisation can implement to be more positive, such as (i) ensuring good leadership, (ii) ensuring a positive organisational culture, (iii) adopting open and transparent communication, (iv) giving and receiving feedback, (v) ensuring equality between all employees in the organisation, (vi) addressing and resolving conflict, (vii) sharing success and (viii) recognising and compensating employees with good behaviour.

Conflict resolution and communication: Conflict is intrinsic to all human relationships, so the leader of an organisation must be able to manage and contribute to resolving them, thus promoting transparent and respectful communication between the various employees. In this sense, a good leader will be able to deal with problems between employees, helping them to understand the different perspectives. On the other hand, a bad leader tends to avoid dealing with problems and conflicts in the organisation, causing the conflict to escalate to higher dimensions. It follows that it is essential for the organisation to address the conflict immediately and not postpone its resolution.

In addition to resolving conflicts quickly, the interviewee highlighted the importance of the leadership of organisations adopting politically astute behaviour (i.e., smiling and showing sympathetic behaviour towards other employees and stakeholders and avoiding interacting with negative people on a daily basis) in order to be able to deal with different types of people.

Table 4.10 highlights the main themes and patterns identified in the conversation with Professor Issam Ghazzawi.

4.1.4 Thematic Analysis of Interview with Prof. Dr. Frank Fialho (Dr. Kahuna)

The interview between the research team and Dr. Frank Fialho took place on 20th September 2023. As with the interviews analysed above, the following table (Table 4.11) provides a brief contextualisation of the interviewee—Dr. Kahuna.

The purpose of the interview, which lasted around an hour and fifteen minutes, was to get to know the perspective of a strategy and marketing professional and how both are relevant to making an organisation more positive (Table 4.12).

The role of the leader in defining a purpose: According to the interviewee, the role of a leader is fundamental in all organisations, as it is the leader who has the ability to drive all aspects of the organisation's culture. Leaders, equipped with the necessary skills and being human beings with well-founded principles and values (i.e., with interpersonal skills, emotional intelligence, empathy, compassion and the ability to get to know others), will be able to lead the organisation towards its success.

In addition, it is important for the leader to promote a specific, clear and simple mission and vision. Thus, the leader must have the ability to spread and translate this message to the group and all stakeholders, so that they know and understand the importance of the role they play and thus contribute to the success of the organisation. Having a purpose and objectives also helps employees to be more responsible and committed to the organisation, thus working towards its success.

If all the organisation's stakeholders are working towards the same thing and the organisational culture is right, they will work their magic. Otherwise, if the leader is not competent to deal with toxicity, it will become even worse.

According to Dr. Kahuna, employees also have a role to play in contributing to a more positive organisation, so they should change their mindset, i.e., considering themselves as "their own person, with their own moral values that guide their actions and behaviour", and therefore be "the change they want to see" in the organisation.

Table 4.10 Themes and Patterns identified—interview Prof. Dr. Issam Ghazzawi

Themes and patterns identified	Quotes to justify this evidence
Organisational culture is a determining factor for the future success of an organisation	Professor Issam Ghazzawi highlighted the importance of the organisational culture disseminated by the leadership team and shared by all employees for the company's competitiveness. In this sense, for him, organisational culture is the main driving force that determines how all employees and stakeholders behave and contribute to the success of the company
Positive leadership will have to be able to identify "bad apples" and Corporate Psychopaths and act accordingly	The leadership team of an organisation has an important role to play in identifying the elements of the organisation that do not contribute to making it more positive at all levels. Thus, a good leader will have to identify the negative elements, i.e., the bad apples, as well as the so-called corporate psychopaths, because if they don't, the organisation and its future competitiveness could be damaged, as determined by the literature review As well as being able to identify them, it's important that they adopt the necessary behaviours so that they (i) change the way they act towards other members or (ii) stop being associated with the organisation
National culture has an impact on creating more negative or positive organisations	Every country/nation has a distinct culture, and it is this cultural difference that makes the world in which we live more interesting. However, the Professor pointed out that there are cultures that contribute to the creation and establishment of more negative or positive organisations. According to his perspective, the United States is able to create more positive organisations (for example, through the way it deals with conflict and the leadership style it adopts) than countries like Portugal or China (which avoid conflict, for example)
Leadership and conflict management	According to the professor, conflict and its resolution are a key element in the creation of positive organisations and a positive leadership style. Organisations, acting under his leadership, should be able to manage existing conflict and resolve it quickly, as failure to do so could have disastrous consequences for an organisation's competitive success

Table 4.11 Interviewee identification—Prof. Dr. Frank Fialho

Name	Age	Provenance	Academic training	Function
Dr. Frank "Stephen" Fialho	–	Hesperia, California	Chaffey College AA Degree, Marketing and Business Administration 1979–1981 University of La Verne Bachelor of Science Degree, Business Administration 1995–2000 University of La Verne Master of Business Administration MBA, Marketing 2000–2002 University of La Verne Doctorate, Organisational Leadership 2018–2022	University Professor, Marketing—University of La Verne

Table 4.12 Main themes discussed during the interview with Prof. Dr. Frank Fialho

Topic under discussion	Topics covered
The role of the leader in defining a purpose	– The leader's values; – The role of employees in contributing to a more positive organisation – Change to a "be the change you want to see" mindset
Organisational culture and the profitability of the organisation	– The problem of the transactional attitude of leaders
Innovation and strategic marketing	– The role of the leader in innovation – The influence of strategic marketing

And although leadership is negative, they should always be guided by their moral values, contributing to "small big changes".

On the other hand, leaders should also adopt a leadership style that serves the interests of their organisation, their employees, their community and other stakeholders involved—servant leadership strategy. They should always bear in mind that they are in this position to serve others and not to make a profit.

Organisational culture and the profitability of the organisation: There are leaders who have a transactional attitude, i.e., their main focus is financial return and they lose sight of the fact that it is necessary to develop and invest in the people around them. By doing this, profit is a natural consequence. And consequently, according to the interviewee, you can't tell if an organisation is negative by the profit it generates.

We should therefore measure the negativity of an organisation through (i) psychological tests to analyse which dimensions of negativity are highlighted within the organisation and (ii) the type of language used, i.e., whether cruel and rude language is being used, for example in emails. According to his perspective, when employees feel that they are not respected and that their ideals are not considered valid, they start to develop negative feelings towards the organisation and other colleagues, and thus demonstrate negative and abusive behaviour towards others.

Innovation and strategic marketing: According to the professor, innovation leads to the competitiveness of the organisation and the competitiveness of the nation. However, it is complex to identify innovative products that are relevant to the market in which they operate.

Everyone is innovative, creative and has a good idea, but the role of the leader is to identify those that will solve a problem for the consumer and consequently bring profit to the organisation. Innovation can thus lead to competitiveness, but it is the leader's role to steer it in the right direction, with the organisation's vision in mind. In addition, he or she will also encourage the other members to be innovative and generate ideas that are "the next idea used by the organisation".

For the interviewee, Steve Jobs is a great example of a leader capable of guiding innovation at Apple, as he often saw many good innovative ideas, but they were of no use to the organisation and the vision they wanted to follow. In addition, the product needs to be able to solve a problem—for example, integrating the front camera into the iPhone. However, the organisation must also be able to let go of current products so that new innovative products can be integrated—for example, when Apple discontinued the iPod, one of its best-selling products, in order to incorporate it into the iPhone.

Organisations must therefore be attentive to changes in consumer consumption, always with a solution to a problem in mind. To do this, it is essential that they seek to learn continuously, never having the mindset that they are already "good enough" in their area of activity, because when they do, they stop learning and stop being innovative, and consequently stop being competitive.

According to the professor's perspective, strategic marketing plays an important role in terms of the organisation's innovation and competitiveness. Thus, it should be strategic marketing that leads the organisation's innovation, through the search for new products, new market trends and new consumer habits. However, although strategic marketing is a creative business area, marketers are often too focused on launching the right product at the right time, which can make it difficult to generate "stupid" ideas (Table 4.13).

4.1.5 Thematic Analysis of Interview with Veronica Perez

The interview with Veronica Perez took place on 15th September 2022, with the research team. The purpose of the interview was to understand the interviewee's

Table 4.13 Themes and patterns identified—interview Prof. Dr. Frank Fialho

Themes and patterns identified	Quotes to justify this evidence
The influence of the nation's culture	Every country has its differences, and leaders and managers need to be aware of these differences in order to define a strategy that suits the culture and country in question. Although a strategy may be successful in one culture/country, it may not work in another In addition, according to the professor's perspective, the country's culture can also influence the existence of negative organisations to a greater or lesser extent. So, once again, leaders have the important role of keeping an open mind to cultural differences, in order to be able to observe new perspectives and accept possible differences
The fight against the *status quo*	It is very difficult to fight against the *status quo* and possible negative aspects of the organisation, however, according to the teacher's perspective, leadership has a moral responsibility to contribute to a more positive organisation. On the other hand, all individuals have a responsibility to fight for themselves, set limits and demand to be respected. All employees and leadership must also consciously decide to be more positive
The role of strategic marketing in innovation and new product development	According to the interviewee, strategic marketing should play a key role in innovation and the development of new products. Thus, marketers should be responsible in organisations for identifying possible gaps in the market and therefore developing appropriate solutions However, despite their successful past, organisations cannot be arrogant about their success, as this attitude will make them short-sighted and prevent them from innovating and learning continuously, which will have long-term consequences for their competitive advantage over their competitors

(continued)

Table 4.13 (continued)

Themes and patterns identified	Quotes to justify this evidence
Leaders, the definition of purpose and the influence of moral values	Every organisation needs to define a purpose, a mission and a vision for its future. In this way, both leaders and employees will be able to take responsibility for their work and will also know the purpose of their work and how it will help the organisation achieve its goals In addition, leaders are responsible for creating the organisational culture, which in turn is influenced by the values that govern the leader. The values of empathy, compassion and the ability to understand others are, in the opinion of the teacher interviewed, central to the success and competitiveness of the organisation
Negativity in organisational cultures	There are many negative organisational cultures. However, according to the interviewee's perspective, this type of culture negatively influences the organisation's competitiveness and its continued success. It is therefore important for leaders to have the ability to analyse which type of organisational culture predominates and, if they see negative behaviour, to act accordingly. The professor suggests that leaders gauge the negativity of the organisation through the language shared among employees, i.e., whether they use rude language Despite the leadership's responsibility to contribute to a more positive organisation, it is also the responsibility of each employee to change their mindset, i.e., everyone has the possibility to choose to be positive, as well as to set the necessary limits and demand to be respected by others

position on the role of leadership in an organisation, as a former employee of the United States Air Force and current Assistant Director of Marketing (Table 4.14).

The interview lasted approximately 25 min and Table 4.15 shows the main topics discussed. The topics discussed are explained in more detail below.

Diversity and Competitiveness: Veronica Perez began the discussion by highlighting the importance of diversity for an organisation's competitiveness. According to her, diversity is extremely important because if an organisation follows this principle, it will be able to attract people who are different, both in terms of professional background and personal skills. Diversity stimulates the generation of different ways of seeing the world, so in the organisational world, employees with different ways of thinking and who have had different experiences are essential for the organisation's

Table 4.14 Interviewee identification—Veronica Perez

Name	Provenance	Academic training	Function
Veronica Perez	Fontana, California	San Diego State University BA, Journalism 2009–2014	Assistant Director of Marketing

Table 4.15 Main themes discussed during the interview with Veronica Perez

Topic under discussion	Topics covered
Diversity and competitiveness	– Attracting diverse talent – Importance for stimulating the personal growth of employees and the organisation
Strategic marketing and competitiveness	– Providing a purpose, path and direction to the organisation
The correlation between organisational culture and communication style	– The importance of a transparent and open communication style – The evaluation and feedback system
Negative organisations and their leadership	– Leadership is essential to make change happen – The role of employees in establishing more positive organisations

competitiveness. Therefore, for the interviewee, all organisations should strive for diversity in all aspects. In short, diversity stimulates the growth of employees and the organisation.

Strategic Marketing and Competitiveness: From the interviewee's experience in marketing, strategic marketing is important because it provides the organisation with the direction and path it needs to take in order to achieve the objectives it has set itself. Without a clear strategic marketing plan, the organisation is "drifting", which will ultimately harm its competitiveness and position in the market.

All organisations therefore need clear objectives and a strategic plan to help them achieve them. Veronica Perez also pointed out that a strategic marketing plan will help the organisation to eliminate the "noise" around it, mainly because it is impossible for the organisation to be successful in all areas. Thus, according to the position explained by the interviewee, strategic marketing allows the organisation to be able to identify its objectives, its target audience and why these are relevant to its future, and thus redirect all its actions with these elements as the main focus.

Veronica Perez also pointed out that strategy, and strategic marketing in particular, is important for conveying to the surrounding community everything that the organisation cares about. Without strategy, nothing happens.

In conclusion, each organisation should have its own model, and what works in one organisation may not work in another. However, organisations should keep their overall objectives in mind and therefore understand what they can do to help their employees perform better and more efficiently, so strategy and strategic marketing play a pivotal role in the competitiveness of all types of organisations.

The correlation between organisational culture and communication style: According to the interviewee, organisational culture has a huge impact on employees and, in turn, influences the organisation's competitiveness. In this sense, Veronica

Perez, as mentioned above, highlighted the impact of diversity on the organisation's competitiveness. She also highlighted the role of freedom of expression and innovation within the organisation.

Organisational communication also plays a central role in the organisation's competitiveness, since if the organisation adopts an open communication style, employees will be able to reach their full potential and contribute to the organisation's competitiveness. In short, if employees are guided by the aforementioned values, they will make the team better and more competitive. In addition, the organisational culture should also be able to clearly and directly convey the organisation's mission and objectives in order to provide direction for employees, as they will be able to identify "the reason behind what they do" in their day-to-day work, which also stimulates the organisation's competitiveness, according to the interviewee.

Furthermore, the appraisal system and the way in which feedback is given can change an organisation's competitiveness. For her, all organisations should have a constructive and transparent employee feedback and evaluation system. For example, according to her experience, it is important to have a written report explaining everything related to the appraisee's feedback, so that they can see what they are doing well for the organisation and what aspects could be improved. The written report is also essential so that you can easily access the information later.

The degree of innovation within the organisation is also related to the culture of the organisation, as well as the way communication is carried out, i.e., in order for there to be a spirit of innovation in the organisation, employees should feel encouraged to express themselves freely and communicate transparently.

Negative organisations and their leadership: For Veronica, employees have the power to play a fundamental role in making the organisation more positive. In this sense, the leadership team must listen to its employees in order to define measures that will help the work of both parties. In other words, employees are primarily responsible for spreading the culture and, if there is any negativity in the organisation, it is up to the employee to speak up and alert them to the situation. On the other hand, it is the responsibility of the leadership team to act on this event and to act appropriately.

Employees should do what they can to make the workplace more positive, but they are not capable of making change happen. In this sense, leadership has a central role to play in defining the measures to be taken in the face of these negative events. This should be a collaborative task for which the organisation's leaders are primarily responsible.

Veronica Perez also advised leaders to use metrics that help them understand the degree of negativity in their organisation, such as (i) employee feedback, (ii) employee satisfaction, (iii) employees not feeling intimidated to express their opinions, (iv) employees being happy to come to work, (v) working environment, (vi) leadership style and the degree of trust employees have in the leadership team.

In addition, the interviewee said that the culture of the country could also be a factor in the emergence of more or less negative organisations. If on one side of the globe—the West—there is a concern for the self and personal interests, on the other side of the globe—the East—there is a growing concern for the good of the community. The productivity of each country is also a factor that should be taken into account, since there are countries that are more productive than others. During the interview, the following example was mentioned—organisations could consider setting up a factory in Morocco, as wages are lower compared to other countries, but research shows that it is difficult for organisations to recoup the investment made, as productivity in this country is low (on average employees are only productive for two hours a day).

On the other hand, two companies from two different countries will not have similar objectives, so the way they achieve these objectives and the way they work will differ (Table 4.16).

4.1.6 Thematic Analysis of the Interview with the Anonymous Participant

This subchapter aims to analyse the conversation with the sixth participant in the study, who preferred to remain anonymous. A brief description of the participant can be seen in Table 4.17.

During the conversation with the participant, it was possible to get his perspective on Portugal's competitiveness and the way organisations work. Table 4.18 shows the main topics discussed during the interview.

Innovation in Portugal: According to the participant's perspective, the Portuguese business sector has noticed an interesting phenomenon regarding the innovative capacity of start-up companies. Although smaller companies have less capital available to invest in developing new products or services, it is the larger Portuguese companies that find it very difficult to be innovative. Therefore, for the participant, the traditional Portuguese business sector is characterised by its lack of dynamism, which is detrimental to the country's competitiveness.

In this sense, the participant said that the success of organisations depends fundamentally on innovation. However, entrepreneurship and the founders and managers of start-ups in Portugal face a major challenge to their continuity—a lack of capital. Thus, when these companies need to take the step, for example to mass produce products, they find it difficult to do so because they are generally unable to acquire funding and, as a result, these companies and innovative ideas are sold to large international companies. For the participant, start-ups in Portugal "either die or are sold".

In addition, Portuguese companies, such as EDP, which is considered an innovative company, have problems with functionalisation, i.e., there is an internal culture of creating programmes and new initiatives without often measuring the efficiency of these initiatives and their relevance. The participant highlighted the importance of organisations adopting a SMART approach, i.e., anticipating the return on the initiative. From their perspective, Portuguese organisations are not very focused on the return that the implementation of innovation will generate and end up innovating without a strategic plan and without defined objectives.

Table 4.16 Themes and patterns identified—Veronica Perez

Themes and patterns identified	Quotes to justify this evidence
The role of diversity in an organisation's competitiveness	For Veronica Perez, diversity plays an extremely important role in competitiveness and generating competitive advantage, since only in this way will an organisation be able to leverage its way of working. That said, by being able to attract different individuals with different experiences and backgrounds, it will be possible to establish a spirit of collaboration and healthy discussion within the organisation, which brings great benefits for generating an innovative and competitive organisational environment
The consequences of communication style on competitiveness	In addition to the aforementioned importance of diversity, organisational communication is also a factor that, for Veronica Perez, has consequences for competitiveness. If an organisation communicates transparently with its employees, they will be more committed to its mission. What is more, with an objective and transparent evaluation and feedback system, it will allow all employees to make improvements, both in the way they work and in the skills they need to fulfil their role In this way, the interviewee believes that the organisation will be able to improve continuously and become more competitive in its market
The correlation between the definition of organisational objectives and the implementation of an efficient strategic marketing plan	It is important for organisations to set objectives that are relevant and appropriate to their core business and organisational mission. Subsequently, it is also important to define and implement a strategic marketing plan that helps organisations achieve their objectives and maximum potential
The role of leadership in establishing a more positive and therefore more competitive organisation	Although employees play an important role in highlighting possible problems in the organisation, for the interviewee it is the leadership team that has the power and means to make changes that contribute to a more positive organisation If any unhealthy situation occurs that negatively affects employees, the leadership team should make an assessment and take the appropriate measures. It's important to note that only in this way will employees feel comfortable performing their duties and, consequently, contribute to the organisation's competitiveness

Table 4.17 Interviewee identification—anonymous participant

Name	Age	Provenance	Academic training	Function
Anonymous participant	–	Porto, Portugal	Law Degree—Private University, in Porto, 1997	Senior Partner, Portuguese Law Firm

Table 4.18 Main themes discussed during the interview with the anonymous participant

Topic under discussion	Topics covered
Innovation in Portugal	– Innovation in start-up companies – The lack of capital in Portugal – The importance of measuring the efficiency of the innovations to be implemented
The impact of culture on competitiveness	– The relationship culture present in Portuguese society and the lack of meritocracy – The impact of culture on competitiveness
The competences of Portuguese managers	– The skills of Portuguese entrepreneurs – The importance of resilience

The impact of culture on competitiveness: For the anonymous participant, Portugal has a culture that favours relationships with others, meaning that managers prefer to work with individuals with whom they have an affinity and, consequently, promotions are decided with this principle in mind. Portuguese managers favour loyalty over the person with the most technical and interpersonal skills, which means that this country is not governed by a culture of merit.

The anonymous participant also pointed out that the way the Portuguese perceive the success and wealth of their peers has a negative impact on the competitiveness of organisations and, consequently, on the country's competitiveness. The dichotomy between Portugal and the United States of America stands out, as the latter is known for having a culture of succeeding rooted in the "American Dream" and the myth of the emigrant who comes to the country and triumphs, so entrepreneurs, as well as being seen as capitalists, are also recognised as having succeeded in life. On the other hand, Portugal generally perceives the rich as someone who has stolen, who has inherited the family's wealth or someone who has "stepped on other people to get where they are".

In this sense, and as mentioned above, the cultural characteristics of the countries encourage the emergence of more negative and less competitive organisations. On the other hand, the participant emphasised that there is beginning to be a culture of success in Portugal, even if it is very insipid at the moment. The participant pointed out that there are still contradictory forces and that the mentality of the business community does not encourage innovation (as it is essentially risk-averse).

The skills of Portuguese managers: During the conversation with the participant, he emphasised that another of the problems for the country's competitiveness relates to the skills of entrepreneurs. With a business fabric made up of small and medium-sized enterprises, the entrepreneur is generally a dynamic individual, but with little specific training in the business skills and soft skills of an organisation, which has an impact on its competitiveness.

Therefore, according to the participant, it would be very advantageous for entrepreneurs to opt for training to improve these shortcomings, not only for their organisations, but also for the country as a whole.

In addition, the participant emphasised the importance of resilience for all Portuguese entrepreneurs. In Portugal, according to his point of view, you have to be extremely resilient to be successful, as this is a culture that does not reward success, as it is still a very traditional society, and there is therefore little social ascension and status plays too important a role (Table 4.19).

4.1.7 Focus Group Thematic Analysis

This subchapter aims to analyse the conversation and discussion that took place during the Focus Group. Table 4.20 provides a brief description of the participants, who preferred to remain anonymous (Table 4.21).

The influence of a relationship culture: The culture of the nation has repercussions on the culture of the organisation. In this sense, it is important to understand how this aspect shapes the decisions that top leaders make, as well as the organization's performance.

Participants say that Portugal has a more relational culture, as do countries like Spain and Italy. In this sense, the *status quo* and installed power have a great influence on the decisions made and the way organisations work, especially when compared to countries like Canada or the United States of America—countries with a more transactional culture. Innovating in an organisation with a relational culture becomes a challenge, since the *status quo* tends to expel innovators who threaten the structure despite being competent, efficient and everything the company needs.

On the other hand, top leaders in organisations with relational cultures face an extra challenge—the power of "wedges". Focus group participants revealed that this is a day-to-day problem and that it is not always possible to avoid hiring/promoting an individual who is not always the best person for the job. The participants also revealed that this is a problem that could have serious consequences for competitiveness in the long term.

The creation of appraisal systems: in order to foster a more equal and fair working environment, it is important for organisations to develop appraisal systems suited to their needs.

The participants emphasised that this is an area of great sensitivity, since it is not always easy to implement these systems according to plan. It was also stated that although evaluation systems with a very interesting architecture are developed, it is not always possible to implement such systems due to objections from the leadership itself. Managers and leaders often do not accept the proposed system and end up making changes according to what they think is most convenient, jeopardising

Table 4.19 Themes and patterns identified—anonymous participant interview

Themes and patterns identified	Quotes to justify this evidence
The power of the *status quo* in inhibiting innovation	For the anonymous participant, the incumbent *status quo* has great power when it comes to inhibiting innovation. This is because when there is an organisational culture that neither rewards merit nor recognises it, all innovation tends to be viewed with suspicion, especially in companies where hierarchical structures are watertight, very vertical and not horizontal. So, according to the anonymous participant, the main barrier to innovation is not so much the country's technological backwardness, which is now being recovered, but more precisely the mistrust that exists towards those who innovate in organisations and who bring value
The academic qualifications of the Portuguese and the student loans model	According to the anonymous participant, the qualifications of Portuguese people and entrepreneurs are one of the main barriers to positive organisations and competitiveness. Even though there is a high level of Portuguese acquiring higher education, according to his perspective, the types of courses on offer are not suited to the needs of the market, i.e., there should be more technical courses on offer In addition, the anonymous participant revealed the existence of a "doctoral culture", in which researchers aim to acquire an academic title, without the research being useful to society The anonymous participant also pointed out that a system of student loans should be set up, as this would help universities to channel important resources into research, innovation and the development of new products. Education in Portugal is currently very expensive for the Portuguese state, yet students go abroad to work after completing their studies, as they have a more competitive salary and better living conditions Student loans would make it possible for students to do their degrees, but they would not overburden universities, because state resources currently go towards paying teachers and infrastructure, so there is no money left over to pay for research and to retain students. This measure would therefore boost the competitiveness of the university itself, but also of the country, which in the long term would be able to compete on a global scale and retain its young graduates
The power of teamwork in competitiveness	For the anonymous participant, the ability to work as a team is central to every business and every organisation. By exchanging opinions and sharing tasks, teams have enormous power, enabling the organisation to be better and more competitive
State sponsorship	The anonymous participant emphasised that Portugal is a country that lives on the back of state sponsorship, so he values individuals who want to succeed on their own by being entrepreneurs and starting their own companies

Table 4.20 Identification of Focus Group participants

Name	Age	Provenance	Academic training	Function
Anonymous stakeholder A	–	Porto, Portugal	Ph.D., Department of Industrial Engineering and Management, University of Porto (FEUP)	University Professor
Anonymous stakeholder B	–	Porto, Portugal	Degree in Economics, Faculty of Economics, University of Porto (FEP)	Business consultant
Anonymous stakeholder C	–	Porto, Portugal	Degree in Economics, Faculty of Economics, University of Porto (FEP)	Business consultant

Table 4.21 Main themes discussed in the Focus Group

Topic under discussion	Topics covered
The influence of a relationship culture	– The power of the *status quo*; – The difficulty of innovating in organisations with a relationship culture
The creation of evaluation systems	– The importance of evaluation systems; – The difficulty of implementing good evaluation systems
The importance of motivation for top leaders	– The importance of motivation and talent; – The role of resilience and focus in a leader's success

the whole appraisal model. It was also pointed out that an appraisal system should be made up of objective and measurable metrics, but it should also have some subjective criteria that make it possible to understand how the employee develops their interpersonal relationships.

It was also mentioned that a leader should keep a certain distance from employees. Participants say that nowadays there is the idea that personal life should be shared, but when it means sharing with the leader or someone in a managerial position, this person will not fulfil their role in the best way and their activity as a leader is therefore compromised because at a certain point, after a lot of sharing, the leader already knows everything about the employee's life.

The importance of motivation for top leaders: Participants said that motivation is an essential tool for all leaders, but there is no magic formula for fostering motivation. It was also mentioned that the talent of the individual is an important part, but it is not an essential component for success, but of course it helps in the process.

In addition to motivation, it was also mentioned that resilience and focus play an important role in a leader's success. Just like top athletes, leaders must be resilient in

Table 4.22 Themes and patterns identified in the Focus Group

Themes and patterns identified	Quotes to justify this evidence
Hard skills and soft skills	Participants pointed out that hard skills are very important for a leader's success, as it is essential that they understand the business in depth and are able to take the organisation to the next level. However, it was also emphasised that it is essential for leaders to acquire soft skills that enable them to better communicate their message, as well as to manage those around them Soft skills such as motivation, resilience and focus were highlighted throughout the discussion and it became clear how important they are for success, competitiveness and making the organisation more positive
The power of the *status quo* in generating innovation	For the participants, maintaining the *status quo* has a significant influence on restricting innovation. This phenomenon stems from the existence of an organisational culture that does not value merit or recognition, which results in a general mistrust of innovation, especially in companies with an organisational culture centred on interpersonal relationships. Therefore, according to these participants, the main barrier to innovation lies not so much in the country's technological backwardness, which is in the process of catching up, but precisely in the organisational culture oriented towards interpersonal relationships
The toxicity of the relational culture	In an organisational context where the culture is strongly based on interpersonal relationships, significant challenges can arise. This approach can lead to promotion based on personal relationships, to the detriment of merit, demotivating skilled talent. In addition, a relational culture can resist innovation, fearing that change could upset existing relationships Furthermore, decisions influenced by personal relationships can generate perceptions of inequality and injustice, affecting morale and organisational cohesion

order to overcome the challenges they face and to continue to persist in their goals and work towards them (Table 4.22).

4.2 Discussion of the Results

This chapter aims to highlight the most important ideas that emerged from all the qualitative fieldwork carried out and presented in the previous Sect. (4.1 Results of the interviews and Focus Group) and will also leverage the strategies suggested and listed in the following Sect. (4.3—Competitiveness Strategies—Playbook).

Through the interviews and the Focus Group it was possible to identify common points, as well as points that differed between the various conversations. This chapter will therefore be subdivided into the following three questions:

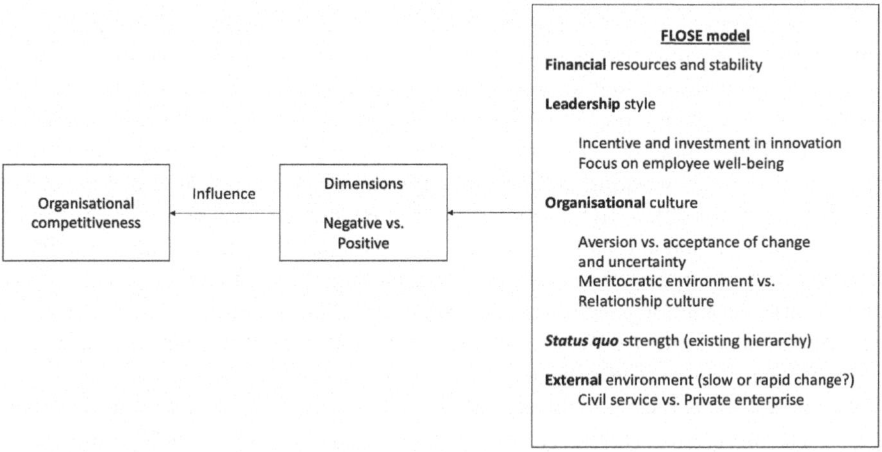

Fig. 4.1 Conceptual model—FLOSE (Own elaboration)

i. Do leadership and organisational culture play a central role in the subject under study?
ii. Is Portugal an incubator for Negative Organisations?
iii. Case study: Is Google, a once very positive firm, losing its competitive edge?

Based on all the theoretical and practical work carried out, the authors developed the conceptual model illustrated in Fig. 4.1. This model therefore aims to explain that the competitiveness of organisations is influenced by various FLOSE factors or dimensions, such as the financial resources and stability of the organisation, the leadership style, the organizational culture, the strength of the *status quo*, and the external environment (characterised by rapid or slow change)—consequently defining whether an organisation is more negative or positive. A more negative organization will experience, in the medium to long run, less financial stability, a poor leadership style which does not provide incentives for innovation nor focus on employee well-being; with an organisational culture which dislikes change and uncertainty, and is more focused on existing relationships than on a meritocracy; if the *status quo* is especially strong and the environment slow to change, then the company risks becoming more negative and thus less competitive.

4.2.1 Do Leadership and Organizational Culture Play a Central Role in the Subject Under Study?

Through the fieldwork carried out, it was possible to see that leadership and organisational culture play a central role in all types of organisations, and therefore have the ability to influence, through their way of acting, whether an organisation is more negative or more positive. In this sense, the role of a leader is complex and requires them to be attentive to everything that goes on in their organization—from the way employees behave towards each other to the consumption habits of their consumers.

Firstly, it is essential that the organisation and its leadership are able to establish the short- medium- and long-term objectives they want to achieve. In this way, it will be easier for the whole organisation, including its employees, to focus their activity on these objectives, working and producing with this purpose in mind. However, it is important to stress that these objectives must be aligned with your organisational mission and core business. Due to the ever-changing world, it is also important for organisations to be agile and open to adapting these objectives, as there are always variables that they cannot predict, such as the introduction of a new disruptive technology by a rival firm. However, uncertain environments require a clear vision of the organisation's future objectives.

Therefore, as the literature review has already shown, organisations must be attentive to the entire surrounding ecosystem in order to be continuously competitive. Focusing on the future and future competitiveness may lead, with greater certainty, the organisation to a successful future. Therefore, adaptability and agility are key competences for their competitiveness, so that they are able to thrive in a dynamic business environment.

In addition, innovation is also an essential component for the competitive advantage of all organisations, which need to know how to adapt to new circumstances. During the fieldwork, examples were given of companies—KODAK, IBM and NOKIA—that developed what is known as "marketing myopia", i.e., these organisations failed to be aware of changes in the ecosystem they were part of, even though they were leaders in their market segment at the time. As well as not being able to anticipate possible changes in the market, their leaders were not agile enough to adapt when these changes became a reality. It can therefore be concluded that organisational culture has a major influence on the competitiveness of organisations, through the impetus it gives to agility and the anticipation of change.

It is very important that the leadership team drives the organisation to take nothing for granted, as well as encouraging continuous learning. To do this, leadership needs to stimulate creativity, innovation and teamwork, which are essential to the organisation's competitiveness. The leader thus has the arduous task of identifying the ideas that actually solve a problem for consumers.

In addition, it was found that it is also essential for a leader to orient their activity according to well-founded interpersonal values, i.e., they should lead with empathy and compassion for all their employees, while at the same time promoting the organisation's mission and vision.

Leading an organisation is an arduous and complex task that demands a lot of its leaders. Therefore, creating a positive organisation is a difficult and complex job since individuals are complex and have formatted opinions about everything. Therefore, for an organisation to be positive, it needs to understand the social processes associated with its organisation as well as the technical and commercial processes. It is therefore essential for leaders to understand their employees and their individual way of seeing the world.

In addition, the interviewees stressed that leaders have a responsibility to identify individuals who are not aligned with the organisation's values and contribute to a negative organisational environment. Thus, after identifying the so-called "bad apples", leaders should adopt a firm approach that does not allow these behaviours to be perpetuated in the future and have a negative influence on the organisation's competitiveness, for example through the loss of valuable employees who no longer want to see themselves in the organisation.

It is also important to point out that since organisations are made up of individuals, it is only natural that there is an associated dimension of conflict. However, it is the way in which the organisation and the leader act when faced with conflict that determines the positivity or negativity of the organisation. Organisations should therefore resolve all conflicts openly, so that all parties involved are heard and feel that they are a fundamental piece of the puzzle that is the organisation.

In short, it is possible to see that the organisational culture spread by the leadership team and shared by all employees and other stakeholders has a tremendous influence on the competitiveness of the organisation.

4.2.2 Is Portugal an Incubator for Negative Organizations?

Following the conclusion of the Focus Group, the authors came up with the question that gives this subchapter its name—"Is Portugal an incubator for Negative Organisations?". It turned out that Portugal is indeed an innovative country with a diverse culture that favours the introduction of new products in the market.

However, it was also found that Portugal has a number of complex problems that hinder the development of more positive organisations, and consequently hinder their continued long-term competitiveness in the global market. These problems relate to (i) the enormous tax burden on Portuguese companies, (ii) the skills gap of Portuguese managers, (iii) the inadequacy of Portuguese skills to meet market needs, (iv) the power of the *status quo* and (v) the lack of capital for Portuguese companies, particularly start-ups.

Therefore, although Portugal is an innovative country, it faces a complex landscape of challenges that directly impact the competitiveness of its organisations, and in turn, the competitiveness of the country. As these challenges contribute to a growing number of negative organisations, they face difficulties in thriving and maintaining their competitiveness in a fierce global market.

The high tax burden on Portuguese companies is the basis of the challenge, as entrepreneurs have fewer monetary resources to invest in their organisation and adopt a long-term approach that will allow them to be more competitive. In addition, Portuguese companies have fewer resources to invest in their employees, which means that they are unable to compete with the global market, leading to a loss of valuable national employees.

In addition, entrepreneurs also have gaps in their management qualifications and interpersonal skills (which do not enable them to be a true and effective leader). As a result, they are not prepared to manage their organisations in today's constantly changing and highly complex environment. In addition, the Portuguese population also has gaps in terms of the adequacy of educational qualifications, i.e., although there is a large percentage of the population with higher education, this is often not suited to the real needs of the market. In this sense, the skills gap of entrepreneurs combined with the inadequacy of the population represents the lack of preparation of Portuguese organisations to face the challenges of the future, making them more negative and less competitive. Perhaps more practical VET-type training (vocational education and training) needs also to be a focus in Portugal and by Portuguese entrepreneurs and managers/leaders. Often shorter courses not focused on a higher education degree diploma may better suit market needs (and be more interesting) than courses which are several years long and often only change to adapt/suit market changes every half-a-dozen years. On the other hand, certain research degrees (e.g., at the Ph.D. level) may be rewarding and reap fruits for certain leaders who seek an intellectual challenge up and above day-to-day work and which will bring them to the next intellectual level. There are already certain more practical doctorates in the market, linked to firm projects, which are satisfying the stakeholders involved at all levels.

The power of the incumbent *status quo* and resistance to change are another major challenge facing Portuguese organisations. Portuguese organisations therefore find it difficult to promote innovation and innovators, as there is a huge aversion to risk. As the analysis found, there is no competitiveness without innovation, so it is very difficult for Portuguese organisations to succeed on the international stage when they see innovation and the unknown as something negative and that only "wastes money". Therefore, being highly change averse [22] and having a relationship culture [23] both contribute to a special breeding ground for negative organisations in Portugal. Previous literature supports this [5, 24].

For all the challenges mentioned above, there is one challenge that is a cause-consequence of this panorama—the lack of capital for Portuguese companies, particularly start-ups. Since Portuguese organisations have very limited resources, when the time comes to take the next step towards growth, they face an impasse. They often have to sell their organisation to foreign companies that have the financial resources to invest (in taking the company to the next level) or simply cease their activity.

In short, these problems represent structural problems that most Portuguese organisations face, contributing to them being more negative than positive, and therefore unable to compete and act in the global market. Hence, understanding these challenges and addressing them effectively is essential for transforming these organisations into more positive and competitive entities on the world stage.

4.2.3 Is Google Losing Its Competitive Edge?

When Google was recognised in 2017 as the best company to work for [8] it was thus considered, in the light of the analysis developed throughout this study, to be a positive and competitive organisation. However, it has also been also possible, through the interviews conducted, to see that this position may be in decline.

Google had an organisational culture that set it apart from all other technology companies and had an extraordinary organisational philosophy, which included principles such as "democracy in the works of the web" and "you can make money without doing evil". In addition, the benefits offered by the organisation and the entire support network it provides make possible that its employees also stand out.

However, this research, as presented in Sect. 4.1.2, found that Google's position as a positive, innovative and competitive organisation could be at stake. Its strong and distinctive organisational culture was the main factor contributing to its competitive advantage, promoting innovators and encouraging creativity throughout the organisation. In addition, the transparency of its work and policies was also a crucial factor for the founders.

Google is currently stagnating and in a reactive rather than proactive position, particularly with regard to Chat GPT (and other similar rival technologies). This disruptive new technology represents a threat to Google's future. Albeit what has contributed to Google reaching this point? The analysis carried out showed that its "new" organisational culture and the way creativity is currently viewed represent a major threat to Google's current competitive position.

Therefore, there is a growing attractiveness of Google for more senior employees with greater professional experience who, due to their advanced age, do not want to invest in innovative projects that may never be implemented due to the growing difficulty in implementing new projects in the organisation. On the other hand, the perceived decline in Google's concern with human resources, embodied by its collective dismissal discussed herein, has generated, both inside and outside the organisation, a feeling of disbelief in the management team and the organisation in general.

In short, if Google is unable to re-establish a distinctive organisational culture—based on pillars such as Innovation and Employee wellbeing—it could lose the competitive advantage it currently has. This competitive advantage is under great threat, because at the moment it is only the market leader because there is no competitor that can stand up to it.

4.3 Competitiveness Strategies—Playbook

According to the analysis carried out, and as verified by the literature review, it was possible to see that a negative organisation involves several dimensions and that it is not always easy in practice to identify whether an organisation is positive or negative. Since this analysis involves several dimensions, it is possible for an organisation to be classified as a positive organisation in one dimension, but also negative in another. The process of creating a positive organisation is therefore difficult and complex [2].

However, the qualitative analysis carried out showed that leadership and the organisational culture it disseminates have a major impact on whether an organisation is negative or positive. It is therefore the leadership—the Organisation—that is the most important piece of the puzzle, as it is responsible for the way everything works within the organization—from interpersonal relationships and the way employees are treated to the way innovation is leveraged to make the organisation more competitive. Still, there are two other key pieces to making organisations more positive—the employees and the country's government.

With this in mind, this chapter aims to provide suggestions for strategies that the three key players—the Organisation, the Employees and the Government—can use to their advantage by developing a model that will help them combat the negativity of the organisation.

4.3.1 Strategies for Organizations to Become More Positive

Table 4.23 shows suggestions for strategies that organisations and their leadership can use to make their organisation more positive. This is a complex and time-consuming process, so the organisation needs to take this into account when implementing it.

In addition, this subchapter provides an example of a Strategic Planning Framework that organisations can use to implement the necessary changes [25].

As mentioned above, the following is a Strategic Planning Framework that organisations can use to implement the necessary changes. It is important to note that this methodology does not use sequential timelines, which allows the organisation to adapt the model as it implements the strategies, thus having greater flexibility to respond to situations external to the organisation and which it has no control over [25]. This was the case, for example, with the Covid-19 pandemic. It should also be mentioned that there is no single "one-size-fits-all" model that works for all organisations [8]. There are many other variables, such as the industry in which they operate, so when implementing the following model, organisations must adapt it to their own needs.

4.3.2 Strategies for Employees to Contribute to a More Positive Organization

Employees have a fundamental role to play in making the organisations they work for more positive and competitive. With this in mind and as mentioned above, Table 4.24 shows the strategies that employees can adopt in order to contribute to a more positive organisation.

Table 4.23 Strategies for organisations to become more positive

Area of expertise	Suggested strategies
Innovation	– Invest in continuous innovation, encouraging and promoting Innovation Champions and creativity in all areas of the organisation. In addition, together with the Strategic Marketing department, identify winning ideas that actually solve a problem for the consumer – Develop an innovative mindset in the organisation, enabling them to stimulate habits and routines that allow the organisation to evolve. In order to do this, they must seek continuous training and update preconceived concepts that are already well-founded internally – Create a learning organisation, i.e., in a world of constant change, the whole organisation needs to be prepared to learn continuously. Training on the different tools to be implemented should be encouraged, as well as training that allows the personal development of each employee, for example courses and activities that stimulate the organisation's creativity – Take advantage of new technologies (such as artificial intelligence, autonomous vehicles, big data, 3D printing and blockchain) and understand how the organisation can benefit from using them – Restructure the organisation so that it is agile and flexible—this is the only way it will be able to adapt to continuous change and deliver integrated and innovative solutions to consumers – Create an innovative work culture, fostering a creative culture, for example by removing obstacles such as too many procedures and bureaucracy – Measure innovation and the impact it has on the organisation and its competitiveness
Organisational culture	– Promote an organisational culture based on respect, transparency and cooperation and, consequently, when faced with a negative element (a bad apple) act accordingly and act quickly – Invest in the diversity of the organisational environment, promoting the recruitment and attraction of new talents with diverse backgrounds, professional experiences and personal experiences. It is also important to have people from different cultures. Diversity will therefore foster creativity and the generation of new ideas and perspectives – Create a customer-centric organisational culture by adopting an "Outside In" approach, i.e., listening to the consumer and their needs and then designing a service or product accordingly. It is also crucial to continually revisit the list of products/services offered and adjust it based on the consumer feedback gathered – Create consumer involvement with the organisation, mainly on an emotional level, by designing an entire customer-centred experience – Promote an organisational culture in which employees want to do their jobs well and, moreover, be productive – Provide a sense of trust in employees, creating a safe and supportive space for everyone. It is also essential to encourage courage throughout all hierarchies

(continued)

Table 4.23 (continued)

Area of expertise	Suggested strategies
Leadership	– Adopt a leadership style that best serves all your employees, for example by implementing benefits that allow them to have a better work-life balance – Lead with emotional intelligence, promoting emotional awareness in all hierarchies of the organisation, practising empathy, being vulnerable (showing humility), giving feedback with compassion for others and investing in personal relationships – Address discrimination in the workplace by promoting individual awareness among each employee so that they take into account possible "uncounscious bias". This problem may arise in connection with inclusion and diversity in the organisation, but it is important for the leader and the organisation to be the catalyst for change – Develop efficient teams, defining a clear purpose for everyone in the organisation and therefore having a sense of mission. In addition, it is important to recruit individuals with teamwork skills – Lead for change, i.e., identifying the organisation's needs and what needs to change and welcome change personally, advocating the benefits it will bring to the organisation. Have the motto "be the change you want to see happen" – Develop a positive mindset when faced with change; and thus have the ability to adapt to new processes and circumstances, as well as new ways of thinking – Develop a coaching leadership style, helping employees to grow and succeed, for example by developing a constructive feedback system – Measure and evaluate the degree of negativity in the organisation on a regular basis using metrics such as (i) employee stress, (ii) employee turnover, (iii) employee commitment to the organisation, (iv) pride in belonging to the organisation, (v) employee absenteeism, (vi) employee performance and involvement in the organisation and (vii) inclusion
Management	– Implement and develop an organisational structure that is attentive to the current and future needs of consumers – Create a network of partners that support the organisation's core business, thus subcontracting all activities that are not part of the organisation's core business – Implement a decentralised structure, giving all leaders the opportunity to make decisions within the limits set by the organisation's management team
Communication	– Contribute to the smooth running of remote working, promoting fluid and transparent communication between the different teams and hierarchies, as well as establishing routines and schedules – Be a good communicator, using storytelling, for example, to involve the organisation in a single purpose

(continued)

Table 4.23 (continued)

Area of expertise	Suggested strategies
Recruitment	– Establish a recruitment and promotion process based on the individual's technical and interpersonal skills. Thus, promoting a meritocratic culture in which individuals are not promoted on the basis of relationships with the leadership or acquaintances, but because they add value to the organisation – Ensure that everyone in the organisation learns and develops their skills, thus promoting a culture of quality in the product under development
Organisational well-being	– Create an organisational wellness programme, promoting the creation of organisational wellness programmes for employees, for example physical activity classes. It is also important for the organisation to promote a balance between personal and professional life – Implement a programme to support families, for example by helping with their children's education – Promote the mental health of employees and emphasise the enormous benefits of sleep for productivity and the general well-being of each individual – Promoting a safe and inclusive organisational environment for all employees, allowing them to freely share their opinions and be genuine without fear of reprisals – Empower employees to be more autonomous in their work, trusting them and giving them a purpose and a sense of mission, so that they feel responsible and enjoy working for the organisation

Table 4.24 Strategies for employees to contribute to a more positive organisation

Area of expertise	Suggested strategies
Attitude	– Boost individual resilience by focusing on the things that can be controlled and adopting behaviour that allows you to deal with the stress associated with your professional life, for example by anticipating tasks – Remember the values that guide your behaviour and choose to adopt them, even when faced with a negative person (a "bad apple") – Choose to be positive – Understand reason and purpose, so it is more satisfying to go to work every day. Each employee's attitude defines the value of their work
Change and agility	– Expand your role within the organisation beyond the Job Description, bringing a component of responsibility and motivation to your work – Harness the power of daily habits in your favour, i.e., develop a routine of productive habits allied to your individual goals
Interpersonal relations	– Strengthen positive working relationships in order to bring greater well-being to daily work lives and foster a greater sense of belonging to the team and the organisation
Individual well-being	– Invest in mental and physical health, identifying behaviours that trigger stress, following a physical exercise routine and optimising the balance between personal and professional life

Table 4.25 Strategies for governments to leverage more positive organisations

Area of expertise	Suggested strategies
Tax incentives	– Define tax incentives for more companies to have their headquarters in the country, on the condition that they recruit national employees, thus also promoting job creation for qualified individuals – Reduce the tax burden on individuals, especially young adults, as this will allow them to continue working and living in the country – Reduce taxes for organisations that demonstrate investment in innovation and the development of new products
Retaining young people	– Adopt measures that give incentives to young people not to emigrate, such as benefits on home loans and credit to spend on administrative expenses related to home ownership (i.e., expenses related to the deed)
Benefits to the population	– Improve the public transport network and make it free for all
Innovation	– Improve public education by introducing new curricular units in the area of entrepreneurship and innovation in secondary education, as well as giving greater prominence to curricular units that stimulate young people's creativity

4.3.3 Strategies for Governments to Empower More Positive Organizations

Finally, Table 4.25 lists the strategies suggested for governments to help organisations become more positive. Just as employees play a central role, governments and in particular the measures they implement have a huge influence on the positivity of organisations and their competitiveness.

4.4 How to Implement Strategies? Change Management

Change is a constant in today's organisational world, so a leader must have the ability to influence and alter their organisation's behaviour in order to adapt to this change. It is therefore important that all information is disseminated in a transparent and concise manner so that all employees better understand the changes that will take place, as well as being an element that contributes to the implementation of this change [26].

In this sense, it can be seen that all changes that take place within the organisation must be managed as a process, because the way in which changes are introduced and implemented in the organisation are central to its success. The process of change or transformation should be managed by a specialist, either internal or external to the organisation, who advises the leadership team [2].

To be successful, the change process must follow the following steps [2]:

STEP 1: It is important to establish a good working relationship and co-operation between the CEO and the advisors of the change or transformation process. The relationship between the two change agents, as well as with the entire leadership team, is fundamental for the entire organisation to accept the change and implement it in its day-to-day work.

STEP 2: Articulate, together with the executive director, the organisation's purpose, goals and objectives. The purpose, goals and objectives must be clear to the entire organisation, regardless of the hierarchy.

STEP 3: It is essential that there is a real and deep understanding on the part of the executive leadership of how and why the concept of Systems Leadership Theory and tools can help form a path that enhances the achievement of objectives. Systems Leadership Theory refers to a set of skills and capabilities that any individual or organisation can use to catalyse, enable and support the process of change, thus combining collaborative leadership, coalition building and systems thinking to mobilise innovation and action across a large, decentralised network.

STEP 4: It is important that the leadership, together with the team managing the change process, carries out a structural review of the work environment, such as understanding (i) what work needs to be done and (ii) how to structure the work. There should therefore be a shared understanding with the team of the change required and what the structure might look like in terms of levels of work, operations, services and support roles.

STEP 5: Carry out an audit of systems and symbols, i.e., an organisational "health check". The audit makes it possible to identify different cultures and their unifying mythologies (stories about the organisation that underpin value judgements).

STEP 6: In this step, the CEO formulates the change process taking into account (i) other business/organisational issues and the current context, (ii) what is the order of change and (iii) who will do what? When carrying out this step, it is important to take into account, for example, the appropriate choice of individuals who will take part in the process, as well as the definition of qualitative and quantitative metrics to measure the progress of the change process.

It is important to emphasise that the following phases depend on the extent and complexity of the change itself and may not be followed in the way that follows, i.e., they may be implemented in parallel or even in reverse order.

STEP 7: Train employees in the changes to be implemented. This training should be administered from the top down and co-lead, i.e., with internal course leaders and an external consultant/trainer. The training thus provides an opportunity to discuss the business plan and important information about how the change programme is part of that plan.

STEP 8: Carry out the restructuring necessary for the organisation to achieve its objectives. The authority needed to carry out the work of each function needs to be articulated using a simple and effective job description format. The success of the process depends on the obvious leadership of the relevant chief executive, with the support of internal and external experts. In this sense, it is essential that the process is not dogmatic or overly bureaucratic, although it must be disciplined, as well as differentiating between the work required for a new role and the capacity of the current post holder.

STEP 9: Designing systems, realising that this is a complex process. It is therefore necessary to (i) create small teams to work on the systems (four or five team members maximum), while receiving input from the other teams.

STEP 10: Dismissal of roles, because despite good proposals for structure and systems, people are often left in roles which do not enable them to do the job effectively, so change programmes slow down, stop or fall into disrepute within the organisation.

STEP 11: Continue training throughout the organisation, as it (i) reinforces the common language, (ii) helps to build a culture (people at all levels can discuss their experiences) and (iii) is a positive symbol of equalisation.

STEP 12: Create, design and implement the system. As it is a lengthy process, it is important to select the right people for the job. Processes need to be put in place to measure the effects of the change programme so that it can be easily monitored.

STEP 13: Regularly review the progress of the change process in order to identify, in a way that is clear to the whole organisation, how the process is evolving and to identify the progress that has been made. This emphasises the need to (i) integrate the whole process, (ii) have good control and audit information and (iii) have good measures in place.

References

1. Hitt, M., Ireland, R., Hoskisson, R.:. Strategic Management: Concepts. Competitiveness and Globalisation, 9th edn. South-Western, Mason, USA (2011)
2. Macdonald, I., Burke, C., Stewart, K.: Systems Leadership: Creating Positive Organisations, 2nd edn. Routledge Publishing (2018)
3. Lafley, A.G., Martin, R.L.: Playing to Win–How Strategy Really Works. Harvard Business Review Press, Boston, USA (2013)
4. Porter, M.: What is Strategy? Harvard Business Review, November-December issue (1996)
5. Au-Yong-Oliveira, M.: Negative Organisations and How to Overcome Them. Fighting to Promote Innovation and Change. Sílabas & Desafios, Faro (2022)
6. McGrath, R.G.: The end of competitive advantages: how to keep your strategy to move as fast as your business. Harvard Business Review (cited August 2013). Available from: https://hbr.org/2013/08/the-end-of-competitive-advanta
7. Au-Yong-Oliveira, M., Lebre, I.A.P.M., Nogueira, A.R., Gonçalves, R.: Success, failure, marketing and innovation: the case of Nokia. RISTI-Iber. J. Inf. Syst. Technol. 219–234 (2020)
8. Hsieh, N.H., Klopfenstein, A., Mehta, S.: Culture at Google. Harvard Business Review (2020)
9. Mitchell, T.R., Larson, J.R., Jr.: People in Organisations: An Introduction to Organizational Behaviour, 3rd edn. McGraw-Hill (1987)

10. Porter, M.: The Competitive Advantages of Nations pp. 73–91. Harvard Business Review, March-April (1990)
11. Tse, T.C., Esposito, M.: Understanding How the Future Unfolds: Using DRIVE to Harness the Power of Today's Megatrends. Lioncrest (2017)
12. Hofstede, G., Hofstede, G.J., Minkov, M.: Cultures and Organisations: Software of the Mind. McGraw-Hill (2010)
13. Davey, L.: The status quo is risky, too. Harvard Business Review (cited May 2014). Available from: https://hbr.org/2014/05/the-status-quo-is-risky-too on 9 June 2023
14. Boddy, C.R.: Corporate psychopaths, bullying and unfair supervision in the workplace. J. Bus. Ethics **100**, 367–379 (2010)
15. Boddy, C.R., Taplin, R.: The influence of corporate psychopaths on job satisfaction and its determinants. Int. J. Manpow. 965–988 (2016)
16. Moore, J.F.: Predators and prey: a new ecology of competition. Harv. Bus. Rev. **71**(3), 75–86 (1993)
17. Gulati, V., Seppala, E.: Shifting your perception after being laid off. Harvard Business Review. Available from: https://hbr.org/2023/03/shifting-your-perception-after-being-laid-off in June 2023 on 9 June 2023
18. Google Careers.: Benefits at Google. Available from: https://careers.google.com/benefits/ on 1 June 2023
19. Oliveira, M.: Building leadership: how pride in your work leads to better attendance and quality. Perform. Improv. **44**(7), 5–7 (2005)
20. Pfeffer, J.: Building Sustainable Organisations: The Human Factor. Academy of Management Perspectives (2010)
21. Google Help.: Where You Can Use Bard. Available from: https://support.google.com/bard/answer/13575153?hl=en on 5 June 2023
22. Hofstede, G.: Culture's Consequences—Comparing Values, Behaviours, Institutions, and Organizations across Nations, 2nd edn. Sage publications, Thousand Oaks (2001)
23. Solomon, C.M., Schell, M.S.: Managing Across Cultures–The Seven Keys to Doing Business with a Global Mindset. McGraw-Hill, USA (2009)
24. Au-Yong-Oliveira, M.: Negative organisations and (negative) powerful relationships and how they work against innovation-perspectives from millennials generation Z and other experts. Sustainability **14**, 17018 (2022). https://doi.org/10.3390/su142417018
25. Webb, A.: How to do strategic planning like a futurist. Harvard Business Review. Available from https://hbr.org/2019/07/how-to-do-strategic-planning-like-a-futurist (2019)
26. Simões, P., Esposito, M.: Improving change management: how communication nature influences resistance to change. HAL OPEN SCIENCE. https://hal.science/hal-00802173

Chapter 5
Conclusion

Abstract This chapter concludes the research study and discusses the complex factors that make up either a negative or a positive organisation. Human nature and organizational culture are considered, as is leadership and innovation. Additionally, a final word is spoken on management and managers. The functions of senior management and of middle management differ and are placed in context in this chapter. Middle management is indeed crucial to the well-being of an enterprise, and much can go wrong if it is not productive, efficient and effective. Organisations are political entities and much occurs which is not readily accessible and understandable by participants and passers-by/observers.

The work of an organisation, the way it operates, its employees, its leadership and all the other factors that make it up constitute a set of variables that make them complex.

Judging whether an organisation is positive or negative is also a difficult process, as it involves, as mentioned above, the different dimensions of the company and the human nature of the employees who work there. Even after carrying out this research, there are still questions about whether an organisation is 100% negative or positive, given the large number of dimensions that influence the structure of the organisation. Even so, it was possible to objectively, concisely, and impartially define some criteria that make it possible to identify these dimensions and judge the negativity of an organisation more or less expeditiously.

Organisational culture is one of the pillars by which the negativity of the organisation can be measured. This means that encouraging an organisational culture based on the values of empathy, understanding, respect and transparency, without losing track of the organisation's objectives, is one of the foundations for having a more positive and, consequently, more competitive organisation. Removing or discouraging so-called bad apples or corporate psychopaths is another tangible measure that can be implemented to make the organisation more positive.

Leadership is also the foundation on which the structure of the organisation rests and which can be decisive in defining whether it is negative or positive. It is leaders who approve and stimulate all changes within the organisational fabric, thereby accentuating or attenuating its stagnation or growth towards positivity. After all,

B. Graça and M. Au-Yong-Oliveira, *Competitiveness Strategies for Negative Organizations*, Information Systems Engineering and Management 1,
https://doi.org/10.1007/978-3-031-59914-9_5

leaders have the power to stimulate innovation, change the organisational culture, restructure the organisation's objectives and, ultimately, lead the organisation towards its defined purpose and mission.

Innovation was another crucial aspect that was identified at an early stage as one of the standards for measuring an organisation's negativity. In fact, and taking the above research into account, innovation is not only a driving force to make organisations positive, but also acts as an agent of change within negative organisations. Furthermore, fostering an environment among employees that allows for innovation, initiative, and creativity, elevating those who do it naturally, is another aspect that will allow organisations to challenge the paradigm and *status quo* in which they find themselves.

What is more, as the analysis shows, employees and the government can play a guiding role in the direction the organisation should take. In fact, employees are the working class within the organisation and therefore the ones who emerge as drivers of change when properly demanded. The government, for its part, can boost the emergence of positive organisations by implementing a set of measures that benefit them (please see Table 4.25).

Combining all these aspects is still no guarantee that the company will be competitive. However, combining all these elements will facilitate the company's competitive growth and, consequently, its positivity. Additionally, implementing the above strategies could further boost the company's competitiveness and the way it operates in its market.

There is still a long way to go in the process of metamorphosis, which will be to turn many Portuguese organisations into companies or organisations that assert themselves as being positive, competitive, and innovative organisations. However, it is the authors' perspective that by implementing the strategies defined and with an open mind and an eagerness for change, many Portuguese organisations can naturally emerge as leaders in global markets.

5.1 A Final Word on Management and Managers

We seem to currently be surrounded by negative organizations, on all sides. Who is responsible for this? Top, or senior, management? Middle management? Or lower-level employees with no management responsibilities?

By definition, the managers, both middle and senior, should be responsible for what goes on in an enterprise, and to not assume this is to withdraw authority and decision-making power from the managers. Do we need managers? We do. No doubt. Albeit they have to do their job right.

What is the function of senior management in an enterprise?

Senior management needs to determine the strategic direction to be followed. Lafley and Martin [1] portray this very well in their book *Playing to win—How strategy really works*, and with their cascade of choices which involves addressing and answering five essential questions: (1) What do we want to be (how do we envision ourselves in the future, if all goes according to [our best] plan?)? (2) Where do we want to compete (in what segments? In what geographies? How vertical do we want to be?)? (3) How will we achieve victory [with what competitive advantage?]? (4) What competencies and resources [capabilities] will we need to win? [How will we pair and configure our competencies and resources?] (5) What management (and information) systems will we need in order to support our success?

Management and especially senior management are about making the right integrated choices [1]. To strengthen the brand, target the right consumers, invest in the right innovative products, and proceed in the right direction in view of the tough and growing competition.

Senior management, in enterprises both big and small, will determine where we will invest, where the capital will be invested, and may surround themselves with the best consultants in the industry to help them make the best strategic choices.

As senior management answers directly to shareholders it may be intuitive that they make the best decisions for the company to flourish—and so that they make keep their jobs. Albeit if the shareholders are dispersed this means more power resides with senior management. They will not be so "controlled". On the other hand, if there are shareholders, or certain stakeholders, with a lot of power, then this may limit the strategic path to be followed by the enterprise. One such example that comes to mind is how the most prominent Portuguese private bank controlled and dictated the destiny of PT—Portugal Telecom. By signing away close to a billion euros of their savings to the private bank's own interests, PT failed and soon disappeared. Middle management and the lower-level employees had nothing to do with this demise—it may be solely attributed to senior [miss] management.

As regards middle management, the Japanese authors [2] attribute great importance to middle management. They are the knowledge engineers and catalysts. Their task is one of communication and of execution. They have to implement the plan which was defined by senior management. Albeit one may recognize quite quickly that senior management will fail terribly with poor middle managers in place. Senior managers are often also salespeople, representing the firm and gaining new business, at the highest level. They cannot and should not become too involved in day-to-day operations, as that is not their job—to implement. Or is it? Managers have to delegate, at the end of the day, especially senior managers. Or, otherwise, risk having no private/personal life to speak of outside work. To do senior management's job as well as intermediate management's job is a lot of work. Almost impossible. Albeit if middle management fails outright, they will have to be replaced (is that possible in the case of civil servants, for example?) or senior management will have to do their job for them—if they have access to the information required. Hence the importance of information and management systems—to make possible and accessible to anyone the execution aspect of the business or enterprise.

An example of how middle managers are important and how their behaviour may lead to failure is when:

(1) They are battling to be promoted and hence may follow orders from above without questioning them. This is very dangerous. Please note that questioning is not the same as disobeying. To question, with diplomacy, and while saving the face of senior managers, with a lot of social and soft skills, in the process, is very important. The organizational culture should also be open to this sort of questioning by middle management.
(2) Middle management has lost hope of promotion with the current senior management team and hence is set on sabotaging their work—either by doing nothing (the lesser of the two evils) or by doing things wrong on purpose (this may lead very quickly to disastrous results—except perhaps in the civil service).
(3) Middle management may simply be incompetent and may thus not be able to execute or follow-through on orders from above. This means that middle management may need to be replaced or trained to gain new competencies.

References

1. Lafley, A.G., Martin, R.L.: Playing to Win–How Strategy Really Works. Harvard Business Review Press, Boston, USA (2013)
2. Nonaka, I., Takeuchi, H.: The knowledge-creating company. In: How Japanese Companies Create the Dynamics of Innovation. Oxford University Press, New York, USA (1995)

Appendix A

PRISMA 2020 main checklist

Topic	No	Item	Location where item is reported
Title			
Title	1	Identify the report as a systematic review	N.A
Abstract			
Abstract	2	See the PRISMA 2020 for Abstracts checklist	
Introduction			
Rationale	3	Describe the rationale for the review in the context of existing knowledge	2.2.1
Objectives	4	Provide an explicit statement of the objective(s) or question(s) the review addresses	2.2.1
Methods			
Eligibility criteria	5	Specify the inclusion and exclusion criteria for the review and how studies were grouped for the syntheses	2.2.1
Information sources	6	Specify all databases, registers, websites, organisations, reference lists and other sources searched or consulted to identify studies. Specify the date when each source was last searched or consulted	2.2.1

(continued)

B. Graça and M. Au-Yong-Oliveira, *Competitiveness Strategies for Negative Organizations*, Information Systems Engineering and Management 1,
https://doi.org/10.1007/978-3-031-59914-9

(continued)

Topic	No	Item	Location where item is reported
Search strategy	7	Present the full search strategies for all databases, registers and websites, including any filters and limits used	2.2.1
Selection process	8	Specify the methods used to decide whether a study met the inclusion criteria of the review, including how many reviewers screened each record and each report retrieved, whether they worked independently, and if applicable, details of automation tools used in the process	2.2.1
Data collection process	9	Specify the methods used to collect data from reports, including how many reviewers collected data from each report, whether they worked independently, any processes for obtaining or confirming data from study investigators, and if applicable, details of automation tools used in the process	2.2.1
Data items	10a	List and define all outcomes for which data were sought. Specify whether all results that were compatible with each outcome domain in each study were sought (e.g. for all measures, time points, analyses), and if not, the methods used to decide which results to collect	2.2.1

(continued)

(continued)

Topic	No	Item	Location where item is reported
	10b	List and define all other variables for which data were sought (e.g. participant and intervention characteristics, funding sources). Describe any assumptions made about any missing or unclear information	2.2.1
Study risk of bias assessment	11	Specify the methods used to assess risk of bias in the included studies, including details of the tool(s) used, how many reviewers assessed each study and whether they worked independently, and if applicable, details of automation tools used in the process	N.A
Effect measures	12	Specify for each outcome the effect measure(s) (e.g. risk ratio, mean difference) used in the synthesis or presentation of results	N.A
Synthesis methods	13a	Describe the processes used to decide which studies were eligible for each synthesis (e.g. tabulating the study intervention characteristics and comparing against the planned groups for each synthesis (item 5))	2.2.1
	13b	Describe any methods required to prepare the data for presentation or synthesis, such as handling of missing summary statistics, or data conversions	N.A
	13c	Describe any methods used to tabulate or visually display results of individual studies and syntheses	2.2.1

(continued)

(continued)

Topic	No	Item	Location where item is reported
	13d	Describe any methods used to synthesise results and provide a rationale for the choice(s). If meta-analysis was performed, describe the model(s), method(s) to identify the presence and extent of statistical heterogeneity, and software package(s) used	2.2.1
	13e	Describe any methods used to explore possible causes of heterogeneity among study results (e.g. subgroup analysis, meta-regression)	N.A
	13f	Describe any sensitivity analyses conducted to assess robustness of the synthesised results	N.A
Reporting bias assessment	14	Describe any methods used to assess risk of bias due to missing results in a synthesis (arising from reporting biases)	2.2.1
Certainty assessment	15	Describe any methods used to assess certainty (or confidence) in the body of evidence for an outcome	N.A
Results			
Study selection	16a	Describe the results of the search and selection process, from the number of records identified in the search to the number of studies included in the review, ideally using a flow diagram	2.2.1
	16b	Cite studies that might appear to meet the inclusion criteria, but which were excluded, and explain why they were excluded	2.2.1
Study characteristics	17	Cite each included study and present its characteristics	Bibliographical references
Risk of bias in studies	18	Present assessments of risk of bias for each included study	N.A

(continued)

(continued)

Topic	No	Item	Location where item is reported
Results of individual studies	19	For all outcomes, present, for each study: (a) summary statistics for each group (where appropriate) and (b) an effect estimate and its precision (e.g. confidence/credible interval), ideally using structured tables or plots	N.A
Results of syntheses	20a	For each synthesis, briefly summarise the characteristics and risk of bias among contributing studies	N.A
	20b	Present results of all statistical syntheses conducted. If meta-analysis was done, present for each the summary estimate and its precision (e.g. confidence/credible interval) and measures of statistical heterogeneity. If comparing groups, describe the direction of the effect	N.A
	20c	Present results of all investigations of possible causes of heterogeneity among study results	2.2.1
	20d	Present results of all sensitivity analyses conducted to assess the robustness of the synthesised results	N.A
Reporting biases	21	Present assessments of risk of bias due to missing results (arising from reporting biases) for each synthesis assessed	N.A
Certainty of evidence	22	Present assessments of certainty (or confidence) in the body of evidence for each outcome assessed	N.A
Discussion			
Discussion	23a	Provide a general interpretation of the results in the context of other evidence	1
	23b	Discuss any limitations of the evidence included in the review	N.A

(continued)

(continued)

Topic	No	Item	Location where item is reported
	23c	Discuss any limitations of the review processes used	N.A
	23d	Discuss implications of the results for practice, policy, and future research	Conclusion
Other information			
Registration and protocol	24a	Provide registration information for the review, including register name and registration number, or state that the review was not registered	N.A
	24b	Indicate where the review protocol can be accessed, or state that a protocol was not prepared	N.A
	24c	Describe and explain any amendments to information provided at registration or in the protocol	N.A
Support	25	Describe sources of financial or non-financial support for the review, and the role of the funders or sponsors in the review	N.A
Competing interests	26	Declare any competing interests of review authors	N.A
Availability of data, code and other materials	27	Report which of the following are publicly available and where they can be found: template data collection forms; data extracted from included studies; data used for all analyses; analytic code; any other materials used in the review	2.2.1

Appendix B

PRISMA abstract checklist

Topic	No	Item	Reported?
Title			
Title	1	Identify the report as a systematic review	No
Background			
Objectives	2	Provide an explicit statement of the main objective(s) or question(s) the review addresses	Yes
Methods			
Eligibility criteria	3	Specify the inclusion and exclusion criteria for the review	No
Information sources	4	Specify the information sources (e.g. databases, registers) used to identify studies and the date when each was last searched	Yes
Risk of bias	5	Specify the methods used to assess risk of bias in the included studies	No
Synthesis of results	6	Specify the methods used to present and synthesise results	No
Results			
Included studies	7	Give the total number of included studies and participants and summarise relevant characteristics of studies	No
Synthesis of results	8	Present results for main outcomes, preferably indicating the number of included studies and participants for each. If meta-analysis was done, report the summary estimate and confidence/credible interval. If comparing groups, indicate the direction of the effect (i.e. which group is favoured)	Yes

(continued)

B. Graça and M. Au-Yong-Oliveira, *Competitiveness Strategies for Negative Organizations*, Information Systems Engineering and Management 1,
https://doi.org/10.1007/978-3-031-59914-9

(continued)

Topic	No	Item	Reported?
Discussion			
Limitations of evidence	9	Provide a brief summary of the limitations of the evidence included in the review (e.g. study risk of bias, inconsistency and imprecision)	Yes
Interpretation	10	Provide a general interpretation of the results and important implications	Yes
Other			
Funding	11	Specify the primary source of funding for the review	No
Registration	12	Provide the register name and registration number	No

Appendix C

PRISMA 2020 flow diagram for new systematic reviews which included searches of databases, registers and other sources

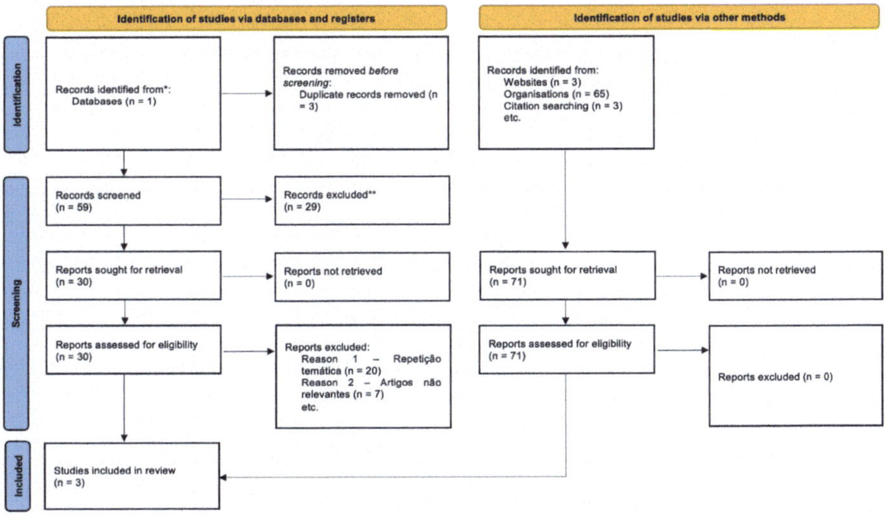

B. Graça and M. Au-Yong-Oliveira, *Competitiveness Strategies for Negative Organizations*, Information Systems Engineering and Management 1,
https://doi.org/10.1007/978-3-031-59914-9

Bibliography

1. Allen, T.: The pandemic is changing employee benefits. Harvard Business Review (cited April 2021). Available from: https://hbr.org/2021/04/the-pandemic-is-changing-employee-benefits
2. Amabile, T.M.: Motivating creativity in organisations: on doing what you love and loving what you do. Calif. Manage. Rev. **40**(1), 39–58 (1997)
3. Amabile, T.M.: How to kill creativity. Harvard Business Review (cited September–October Issue). Available from: https://hbr.org/1998/09/how-to-kill-creativity
4. Arnaout, B., Esposito, M.: The value of communication in turbulent environments: how SMEs manage change successfully in unstable surroundings. Int. J. Entrep. Small Bus. 500–515 (2018)
5. Arranz, N., Arroyabe, M., Li, J., Fernandez de Arroyabe, J.C.: Innovation as a driver of eco-innovation in the firm: an approach from the dynamic capabilities theory. Bus. Strateg. Environ. **29**(3), 1494–1503 (2020)
6. Au-Yong Oliveira, M., Ferreira, J.J.: Business Narrative Modelling Language (BNML) representations: a storyline view and a plot view of how interoperability is an antecedent of innovation at IP BRICK. Afr. J. Bus. Manag. 5338–5361 (2012)
7. Au-Yong-Oliveira, M.: Negative Organisations and How to Overcome Them. Fighting to Promote Innovation and Change. Sílabas & Desafios, Faro (2022)
8. Au-Yong-Oliveira, M.: Negative organisations and (negative) powerful relationships and how they work against innovation-perspectives from millennials generation Z and other experts. Sustainability **14**, 17018 (2022). https://doi.org/10.3390/su142417018
9. Au-Yong-Oliveira, M., Lebre, I.A.P.M., Nogueira, A.R., Gonçalves, R.: Success, failure, marketing and innovation: the case of Nokia. RISTI-Iber. J. Inf. Syst. Technol. 219–234 (2020)
10. Au-Yong-Oliveira, M., Marinheiro, M., Tavares, J.: The power of digitalisation: the Netflix story. WorldCIST. **2**, 590–599 (2020)
11. Beck, L., Trombetta, W., Share, S.: Using focus group sessions before decisions are made. Med. J. **47**(2), 73–74 (1986)
12. Bell, J.: How to carry out a research project-a guide to research in the social sciences and education, 5th edn. Gradiva (2010)

B. Graça and M. Au-Yong-Oliveira, *Competitiveness Strategies for Negative Organizations*, Information Systems Engineering and Management 1,
https://doi.org/10.1007/978-3-031-59914-9

13. Boddy, C.R.: Corporate psychopaths, bullying and unfair supervision in the workplace. J. Bus. Ethics **100**, 367–379 (2010)
14. Boddy, C.R., Taplin, R.: The influence of corporate psychopaths on job satisfaction and its determinants. Int. J. Manpow. 965–988 (2016)
15. Breidenthal, A.P., Liu, D., Bai, Y., Mao, Y.: The dark side of creativity: coworker envy and ostracism as a response to employee creativity. Organ. Behav. Hum. Decis. Process. **161**, 242–254 (2020)
16. Cai, W., Li, G.: The drivers of eco-innovation and its impact on performance: evidence from China. J. Clean. Prod. **176**, 110–118 (2018)
17. Comparably.: Google. Available from: https://www.comparably.com/companies/google in June 2023
18. Chesbrough, H.: The Era of Open Innovation. MIT Sloan Management Review, Spring issue (2003)
19. Christensen, C.M., Raynor, M.E.: The Innovator's Solution: Creating and Sustaining Successful Growth. Harvard Business School Press, Boston (2003)
20. Creswell, J.W.: Research Design: Qualitative, Quantitative and Mixed Methods, 2nd edn. ARTMED Editora S.A. (2007)
21. Davey, L.: the status quo is risky, too. Harvard Business Review (cited May 2014). Available from: https://hbr.org/2014/05/the-status-quo-is-risky-too on 9 June 2023
22. Drucker, P.: The theory of the business. Harvard Business Review (cited September-October issue). Available from: https://hbr.org/1994/09/the-theory-of-the-business.
23. Eisner, E.W.: The Enlightened Eye: Qualitative Inquiry and the Enhancement of Educational Practice. Macmillan, New York (1991)
24. Elo, S., Kääriäinen, M., Kanste, O., Pölkki, T., Utriainen, K., Kyngäs, H.: Qualitative content analysis: a focus on trustworthiness. SAGE Open **4**(1), 1–10 (2014)
25. Enkel, E., Gassmann, O., Chesbrough, H.: Open R&D and open innovation: exploring the phenomenon. Blackwell Publishing Ltd. Available from: https://onlinelibrary.wiley.com/doi/abs/, https://doi.org/10.1111/j.1467-9310.2009.00570.x (2009)
26. Esposito, M., Tse, T.: DRIVE: the five megatrends that underpin the future business, social, and economic landscapes. Thunderbird Int. Bus. Rev. **60**(1), January–February, 121–129 (2017)
27. European Commission.: European Innovation Scoreboard 2021-Coutry Profile Portugal. Available from: https://ec.europa.eu/assets/rtd/eis/2023/ec_rtd_eis-country-profile-pt.pdf
28. Fernandes, P.: Corruption: Definition, Prevention and Combat-The Case of S. Tomé and Príncipe. University of Aveiro. Autonomous Section of Social, Legal and Political Sciences. Aveiro (2009)
29. Fernández, Z., Nieto, M.J.: Internationalisation strategy of small and medium-sized family businesses: some influential factors. Fam. Bus. Rev. **18**(1), 77–89 (2005)
30. Franco, M., Matos, P.G.: Leadership styles in SMEs: a mixed-method approach. Int. Entrep. Manag. J. **11**, 425–451 (2015). https://doi.org/10.1007/s11365-013-0283-2
31. Google Careers.: Benefits at Google. Retrieved from: https://careers.google.com/benefits/ on 1 June 2023.
32. Google Help (2023). Where you can use Bard. Available from: https://support.google.com/bard/answer/13575153?hl=en on 5 June 2023
33. Gross, C.: Collective strategy condensation: when envy splits societies. Entropy **23**(2), February, 157, 1–10 (2021)
34. Gonçalves, V., Mendes, F., Sardinha, I., Rodrigues, R.: Twenty years after the Porter Report for Portugal. Compet. Rev. **25**(5), 540–554 (2015)
35. Gulati, V., Seppala, E.:. Shifting your perception after being laid off. Harvard Business Review. Retrieved from: https://hbr.org/2023/03/shifting-your-perception-after-being-laid-off in June 2023 on 9 June 2023
36. Hitt, M., Ireland, R., Hoskisson, R.: Strategic Management: Concepts. Competitiveness and Globalisation, 9th edn. South-Western, Mason, USA (2011)
37. Hofstede, G.: Culture's Consequences-Comparing Values, Behaviours, Institutions, and Organisations Across Nations, 2nd edn. Sage, Thousand Oaks (2001)

38. Hofstede, G., Hofstede, G.J., Minkov, M.: Cultures and Organisations: Software of the Mind. McGraw-Hill (2010)
39. Hofstede Insights Oy.: Hofstede insights-COUNTRY COMPARISON TOOL. Available from: https://www.hofstede-insights.com/country-comparison-tool?countries=portugal in September 2023
40. Howell, J.M.: The right stuff: identifying and developing effective champions of innovation. Acad. Manag. Exec. **19**(2), 108–119 (2005)
41. Hsieh, N.H., Klopfenstein, A., Mehta, S.: Culture at Google. Harvard Business Review (2020)
42. Lafley, A.G., Martin, R.L.: Playing to Win–How Strategy Really Works. Harvard Business Review Press, Boston, USA (2013)
43. Lincoln, Y.S.: The making of a constructivist. In: Guba, E.G. (ed.) The Paradigm Dialogue, pp. 67–87. Sage, Newbury Park, CA (1990)
44. Lubber, M.: Climate change: investors' next global mega-trend? (cited August 2008). Harvard Business Review. Available from: https://hbr.org/2008/08/climate-change-investors-next
45. Machiavelli, N.: The Prince. Dover Publications Inc., New York, USA (2000)
46. Macdonald, I., Burke, C., Stewart, K.: Systems Leadership: Creating Positive Organisations, 2nd edn. Routledge Publishing (2018)
47. Martin, R.L.: Strategy is iteractive prototyping. Harvard Business Review (cited June 2014). Available from: https://hbr.org/2014/06/strategy-is-iterative-prototyping
48. Martin, R.L.: Strategic choices need to be made simultaneously, not sequentially. Harvard Business Review (cited April 2017). Available from: https://hbr.org/2017/04/strategic-choices-need-to-be-made-simultaneously-not-sequentially
49. McGrath, R.G.: The end of competitive advantages: how to keep your strategy to move as fast as your business. Harvard Business Review (cited August 2013). Available from: https://hbr.org/2013/08/the-end-of-competitive-advanta
50. Miles, M.B., Huberman, A.M.: An Expanded Sourcebook: Qualitative Data Analysis, 2nd edn. SAGE Publications, Inc. (1994)
51. Mitchell, T.R., Larson, J.R., Jr.: People in organisations: An introduction to Organizational Behaviour, 3rd edn. McGraw-Hill (1987)
52. Moore, J.F.: Predators and prey: a new ecology of competition. Harv. Bus. Rev. **71**(3), 75–86 (1993)
53. Nielsen, K., Randall, R., Yarker, J., Brenner, S.O.: The effects of transformational leadership on followers' perceived work characteristics and psychological well-being: a longitudinal study. Work Stress. **22**(1), 16–32 (2008). https://doi.org/10.1080/02678370801979430
54. Nonaka, I., Takeuchi, H.: The knowledge-creating company. How Japanese Companies Create the Dynamics of Innovation. Oxford University Press, New York, USA (1995)
55. Oliveira, M.: Building leadership: how pride in your work leads to better attendance and quality. Perform. Improv. **44**(7), 5–7 (2005)
56. Page, M.J., Moher, D., Bossuyt, P.M., Boutron, I., Hoffmann, T.C., Mulrow, C.D., Shamseer, L.M., Tetzlaff, J.M., Akl, E.A., Brennan, S.E., Chou, R., Glanville, J., Grimshaw, J.M., Hróbjartsson, A., Lalu, M.M., Li, T., Loder, E.W., Mayo-Wilson, E., McDonald, S., McGuinness, L.A., Stewart, L.A., Thomas, J., Tricco, A.C., Welch, V.A., Whiting, P., McKenzie, J.E.: PRISMA 2020 Explanation and Elaboration: Updated Guidance and Exemplars for Reporting Systematic Reviews. Research Methods and Reporting (2021)
57. Pfeffer, J.: Building Sustainable Organisations: The Human Factor. Academy of Management Perspectives (2010)
58. Pistrui, J.: How Managers Can See the Future More Clearly. Harvard Business Review (cited October 2015). Available from: https://hbr.org/2015/10/how-managers-can-see-the-future-more-clearly
59. Porter, M.: The Competitive Advantages of Nations, pp. 73–91. Harvard Business Review (cited March–April) (1990)
60. Porter, M.: What is strategy? Harvard Business Review (November-December issue) (1996)
61. Porter, M.E.: Competitive strategy. Meas. Bus. Excellence. Milest. **1**(2), 12–17 (1997)

62. Rego, A. (2007). Personal and Organisational Communication: theory and practice. 1st edition. Edições Sílabo, Lda.
63. Rosenberg, N.: Why do firms do basic research (with their own money?). Res. Policy **19**(2), 165–174 (1990)
64. Saunders, M., Lewis, P., Thornill, A.: Research Methods for Business Students, 8th edn. Pearson Education Limited (2019)
65. Schein, E.H.: Organisational Culture and Leadership, 3rd edn. Jossey-Bass (2004)
66. Simões, P., Esposito M.: Improving Change Management: How Communication Nature Influences Resistance to Change. HAL OPEN SCIENCE. https://hal.science/hal-00802173 (2014)
67. Solomon, C.M., Schell, M.S.: Managing Across Cultures–The Seven Keys to Doing Business with a Global Mindset. McGraw-Hill, USA (2009)
68. Stewart, D.W.: Focus groups. The SAGE Encyclopedia of Educational Research, Measurement, and Evaluation, vol. 2, pp. 687–692. Sage Publications, Thousand Oaks(2018)
69. Strauss, A.L., Corbin, J.: Basics of Qualitative Research: Grounded Theory Procedures and Techniques. Sage, Newbury Park, CA (1990)
70. Taylor, B.: Don't let negativity sink your organisation. Harvard Business Review (cited February 2020). Available from: https://hbr.org/2020/02/dont-let-negativity-sink-your-organization on 15 July 2022
71. Tidd, J., Bessant, J.: Managing Innovation: Integrating Technological, Market and Organisational Change, 7th edn. Wiley (2013)
72. Tidd, J., Bessant, J.: Innovation Management, 5th edn. Wiley & Sons Limited (2015)
73. Tjosvold, D.: The conflict-positive organisation: it depends upon us. J. Organ. Behav. **29**, 19–28 (2008) (2007)
74. Tjosvold, D., Su, F.: Manager as mediator: devoloping a conflict-positive organisation. In: Herrman, M.S. (ed.) The Blackwell Handbook of Mediation: Bridging Theory, Research, and Practice. Wiley & Sons (2006)
75. Toor, S.U.R., Ogunlana, S.: Ineffective leadership: investigating the negative attributes of leaders and organisational neutralizers. Eng. Constr. Archit. Manag. **16**(3), 254–272 (2009)
76. Tse, T.C., Esposito, M.: Understanding How the Future Unfolds: Using DRIVE to Harness the Power of Today's Megatrends. Lioncrest (2017)
77. Tupper, H., Ellis, S.: It's Time to Reimagine Employee Retention. Harvard Business Review (2022)
78. Vickers, M.H.: Towards reducing the harm: workplace bullying as workplace corruption-a critical review. Empl. Responsib. Rights J. **26**, 95–113 (2014)
79. Webb, A.: How to do strategic planning like a futurist. Harvard Business Review. Available from https://hbr.org/2019/07/how-to-do-strategic-planning-like-a-futurist (2019)

The manufacturer's authorised representative in the EU is Springer Nature Customer Service Centre GmbH, Europaplatz 3, 69115 Heidelberg, Germany. If you have any concerns regarding our products, please contact ProductSafety@springernature.com

Printed and bound by CPI Group (UK) Ltd, Croydon, CR0 4YY
30/09/2026
02225607-0004